D1308442

How to Lie with Charts

How to Lie with Charts™

Gerald E. Jones

SYBEX®

San Francisco ■ Paris ■ Düsseldorf ■ Soest

Acquisitions Manager: Kristine Plachy

Developmental Editor: Richard Mills

Editor: Kristen Vanberg-Wolff

Book Designer: Gary Palmatier, Ideas to Images

Production Coordinator: Dave Nash

Indexer: Ted Laux

Cover Designer: Design Site and Archer Design

Cover Illustrator: Paul Lee

SYBEX is a registered trademark of SYBEX Inc.

TRADEMARKS: SYBEX has attempted throughout this book to distinguish proprietary trademarks from descriptive terms by following the capitalization style used by the manufacturer.

Every effort has been made to supply complete and accurate information. However, SYBEX assumes no responsibility for its use, nor for any infringement of the intellectual property rights of third parties which would result from such use.

Copyright © 1995 SYBEX Inc., 2021 Challenger Drive, Alameda, CA 94501. World rights reserved. No part of this publication may be stored in a retrieval system, transmitted, or reproduced in any way, including but not limited to photocopy, photograph, magnetic or other record, without the prior agreement and written permission of the publisher.

Library of Congress Card Number: 95-69860

ISBN: 0-7821-1723-6

Manufactured in the United States of America

10 9 8 7 6 5 4 3 2 1

This book is for ARTIS and my fellow wizards Roy, Ted, Joan, Charlene, Cindy, Sue, Maurice, Ed, Terry, and Tom.

Acknowledgments

MANY dedicated and talented people worked diligently to bring you this book, and I'd like to give some of them special mention here.

The credit for seeing virtue in its concept and the courage for advancing its cause goes to Richard Mills, the astute and always considerate developmental editor at Sybex. I owe Richard a further note of thanks for taking me up on my recommendation for a book designer, my friend and colleague Gary Palmatier (Ideas to Images, Santa Rosa, CA). As far as I'm concerned, Gary and his associate Robaire Ream have outdone themselves this time, and I congratulate them for another highly professional job.

A light but firm touch at the author control panel was exercised with wisdom by editor Kris Vanberg-Wolff.

My agent Matt Wagner of Waterside Productions (Cardiff-by-the-Sea, CA) has never sought recognition in my books, but he deserves my gratitude. He's been an able counselor and a sincere friend.

Publicity for this book was handled by Amy Romanoff and Judy Jigarjian of Sybex and Gary Yoshimura, my friend of long standing, at Stoorza, Ziegaus, Metzger & Boyer, Inc. (Los Angeles, CA).

I've enjoyed the support and patient encouragement of Georja Oumano, world-class stand-up comic and life partner, whose biggest fan couldn't be in the audience some nights because he was shacked up with his word processor.

The following computer software applications were used to create the illustrations in this book: Freelance Graphics for Windows (Lotus Development Corporation), Harvard Graphics for Windows (Software Publishing Corporation), Microsoft PowerPoint, Microsoft Project, and Microsoft Excel for Windows (Microsoft Corporation).

The notion of "Don't Show the Don't Knows" described in Chapter 2 should be attributed to James Rothman and was kindly suggested to me by his son Edward.

The image of the motorcyclist in Figure 3.14 is from Bitfolio Edition 6, © Management Graphics Ltd. 1993 (provided to me on CD-ROM by Management Graphics, Inc., Reading, Berks, UK). The chart is adapted from an actual graphic I saw in a sales brochure. The photograph used in Figure 11.2 was provided by Gary Palmatier.

I am indebted to Yoshi Hatakeyama, CEO and Chairman of Zacta Engineering Corp. (Kawasaki City, Japan), for his advice and counsel on my references to Japanese culture in Chapter 5.

Thanks to Peter Nathan, Manager, Technical Marketing, of The Reynolds Group (City of Industry, CA), for suggesting the un-conventional checkbook example in Chapter 9. Fred Gallegos of the U.S. General Accounting Office first introduced me to the worry of worksheets in his book *Audit and Control of Information Systems,* and his personal example convinces me that at least some G-men still wear white hats.

Andrew Hatoff of Pantone, Inc. (Carlstadt, NJ) kindly sent me an evaluation copy of ColorUP color palette software and the reference *The Desktop Color Book* (Verbum Books, 1992), both of which were valuable to me in preparing Chapter 11. Si Becker, past Director of Engineering for the Society of Motion Picture and Television Engineers (White Plains, NY), was the lab expert I consulted in the anecdote at the conclusion of Chapter 11, and he's been a good sport about letting me retell the story. I owe much of my understanding of the subject of color to consulting engineer John Cool.

Much of the material in this book is based on software development work I did in the early 1980s as supervisor of the ARTIS project at Color Terminals International, a division of Creative Technologies, Inc.

Portions of this book first appeared in a series of magazine articles I did for *Computer Graphics World* magazine (1983-1986) and in my books *Looking Good with Harvard Graphics* (ScottForesman, 1991), *A Guided Tour of Excel* (Sybex, 1994), *Font Secrets and Solutions* (Sybex, 1994), *Harvard Graphics: The Art of Presentation* (Prima, 1992) and *Freelance Graphics: The Art of Presentation* (Prima, 1993).

The names of persons and organizations, as well as the data, used in examples and charts in this book are fictional, except where statistical data was derived from published United States government sources in the public domain (the National Debt curve, for example). Deceptive techniques and mistakes shown in the charts are intended for instructional purposes only and do not describe or depict the work products of any actual person or organization.

Table of Contents

Introduction:
Truth Is the Best Revenge!

IF **you feel a twinge of guilt** as you pick up this book, don't worry—you are among friends. I admit that the title is provocative, promising a tantalizing debasement of moral values, at least in the realm of business intercourse. But don't be ashamed that you are tempted to look behind the peepshow curtain. We have all been there, or wanted to.

Make no mistake: The promise of the title is not false. In these pages can be found the potent means to work serious mischief. Call me an optimist, but I have a better opinion of your motives. I can think of several legitimate—even honorable—reasons for your wanting to know how to lie with charts, and I like to think those are the real reasons I wrote this book for you.

For the moment, then, let's assume that you're not a shameless, unprincipled liar who will stop at nothing in your frenzied scurry to the top of the heap. What is there for you here?

You may have been drawn to this book because you feel, as most of us have at one time or another, that *you* have been lied to. Whether you are a manager being presented with a suspiciously rosy sales forecast or an investor being enticed with a pretty addition to your portfolio, you could be easy prey for seductive chartmakers. Learning their nasty tricks is one way to even the odds, if not the score.

Another indication that you need to get acquainted with the tricks of the liars' trade stems from a deep-seated *fear of lying*. When you are the one poised to present, you don't want to get caught with your pants down (unless that's your deliberate plan). In this litigious era, an overdose of caution might be downright healthy. Therefore, go ye and study the liars, that ye may abjure their ways! Let's posit a Golden Rule of the Information Age:

DON'T SHOW UNTO OTHERS WHAT YOU DON'T WANT SHOWN UNTO YOU.

Beyond your own sense of conscience, this crack in your confidence, there's the yawning chasm of public ignorance. For the most part, we are a society of trusting illiterates, where charts are concerned. And the situation is getting worse. Why?

Like so many complexities of postmodern life, computers are at the root of the problem. Charts have become the *lingua franca* of the information age, and you would think that would be a boon to communication. After all, charts are prettier than dreary tables of numbers. And, these days, those pretty pictures are so easy to make! Gone is the tedium of the careful draftsman, especially now that we have pint-sized computers popping up all over the place in our homes and offices. Thanks to the developers of graphics software, it has become a trivial task to create charts of all kinds—in full color, no less.

These days, if your work routine includes charting, it just doesn't make sense to do without the labor-saving miracle of digital personal computing. And it's probably just as well that charts can be made so effortlessly, because otherwise the only people who would be interested in looking at the numbers would be the privileged few who make the decisions, and that strikes me as decidedly undemocratic!

It's been almost two centuries since Charles Babbage and Lady Ada Lovelace tried to invent the computer using a steam engine, but, all things considered, it has been worth the wait. We are fortunate that electronics technology came along just in time to finish the job, that the nuclear arms race stimulated the development of high-performance computer displays, and that the far-sighted wizards of

Silicon Valley responded so capably to the marketplace demand for computer games from a generation called *X* that is fast maturing into a new breed of video-junkie financial analysts who want their information on the screen, in color, and fast.

Thanks to these happy accidents of history, we can now create charts literally at the press of a button.

But I wonder.

Although I'm not ready to totter off with the likes of my fellow curmudgeons, I am old enough to dimly remember a time when we were not so fortunate. I have a vague recollection of peering out through the bars of my crib to watch my honorable forebears labor mightily with pencil, ruler, and graph paper to eke out a meager living on the dusty plain of a world that had yet to invent cheap electronics. (I always knew I had very special parents.)

As I look back on it, I wonder if there wasn't a greater wisdom in those primitive efforts. Having such crude tools might have forced those early chartmakers into slower thought processes. It is conceivable that they actually pondered carefully the composition—maybe even the content!—of those pathetically simple charts and graphs. Can it be that in their technological poverty they achieved a higher level of consciousness? Did they actually come to grasp the *meaning* of their graphic creations?

As has happened again and again throughout human history, in gaining new knowledge we have had to shed the innocence of the past, which once enfolded us like a protective garment. We now stand naked before our computer monitors (at least until two-way video comes into general use). Empowered as the new gods of cyberspace, we have been granted the ability to devour megabytes of data at a single gulp, digest it in mere microseconds, and spew it forth without further thought as visually stunning color imagery.

My fear, though, is that the wisdom of old has disappeared in that brief, electronic *zap!* that transforms raw data into those pretty pictures. The good news (and there will be plenty of it if the Gaia Effect proves to work as advertised) is that, having generated this visual information so thoughtlessly, we can change it in a blink of the other eye. I harbor the hope, then, that there is just time to shape the wet clay of our understanding about charts before it solidifies into something that future generations will not be able to *grok*.

Now that I have you bound and gagged by virtue of your having paid your hard-earned, devalued dollars to remove this book from the opulent comfort of your favorite superstore, you may be wondering why you should heed my advice on the perils of lying with charts. Well, despite the lessons of my early childhood education, I did not set out from a young age to become a guru to chartmakers. My career, like the other happy accidents of history, seems only in retrospect to have been guided by an unseen hand. And, believe it or not, when I peered out of the bars of that crib and saw my father soldering together his first computer from an assortment of colorful resistors and other tiny components, I did not aspire to join the ranks of the digital generation. As soon as I was old enough to get my hands on his graph paper and wrap my chubby little fingers around a pencil, my thoughts turned not to drawing diagrams but to writing silly stories.

As I matured, I fell victim to the economic reality that our society does not cheerfully subsidize aspiring writers. Like a disenfranchised aristocrat in a bygone age, I was forced to learn a trade. I soon discovered that, not surprisingly, technocrats would pay handsomely for magazine articles and books on technical subjects. The narrower the audience and the more arcane the subject, the higher the fee. Somewhere along the way in my budding career as a techno-apologist, I underwent a Kafkaesque transformation. I awoke one morning and found to my dismay that I had changed into a computer nerd.

The genesis of my evolution was my clients' petulant demands for pretty pictures to accompany their droning speeches. As my deadlines grew ever shorter through the innate impatience of my executive taskmasters, I quite naturally yearned for a technological labor-saving solution. I happened on computer-generated slide graphics at a time when it took a roomful of shiny hardware to crunch the numbers. The price of these air-conditioned electronic behemoths would sink a small conglomerate, but fortunately the technocrats had their corporate checkbooks at the ready, because they lusted after those pictures more intently even than they wanted my carefully chosen, jargon-filled prose.

So I found myself the master of a big juice guzzler called a Genigraphics computer, which took many minutes and sucked half the watts out of downtown Detroit to record a single chart on a color slide, but it was a big improvement. And it marked the beginning of the end of profitability for manufacturers of graph paper.

I noticed even then that a disturbing change was taking place. In the old world of wizened artisans who labored with graph paper and the tools in their tabourets, there was time to reflect. In fact, there was a venerated ritual called a *cradle review* (flashback to my crib!) in which artwork pasted on cardboard panels was examined and approved by the client before the monumental expense was incurred of photographing it on color slide film. But no sooner had this Genigraphics thing warmed up than the cherished cradle ceremony—and all other comforting mechanical rituals—disappeared in that same electronic *zap!* The technocrats cheered, delighting in the blinding speed of our new toy, unaware that during their excitement the god of digital imagery had stolen their understanding of their own charts.

Now that electronic microminiaturization has shrunk the Genigraphics machine to the size of a cuddly small rodent, the sagacity of the pet owners shows no signs of improving. These days, you don't need custody of a million-dollar silicon brain to do the job. Anybody with a credit card who knows the way to the local computer store can command the tools of the chartmaker's craft.

It's time that I spoke out, and strongly: These people are lying to you. They are lying to themselves. And if you are doing it, you're both getting it and giving it back. And the shame of it is that there's so little malice involved. For the most part, people don't know that they are telling shameless lies with their gorgeous, new charts. They have put their faith in the *zap!* Someday they will discover—perhaps not too late for the good of the planet—that their new computer god lacks a crucial skill called human judgment.

This book is offered in the sincere hope that you will save its dear price many times over by avoiding the kinds of graphical gaffes that could end up costing you or your capitalist masters big bucks. But beware. This is potent stuff. An unscrupulous person could work this magic to unfair advantage. If you unwisely choose the sinister bend in your life road, you will find no comfort here. You will be on your own through the Dark Wood of Error. But if you desire instead to protect yourself from ruthless liars who somehow manage to lure you into their audience, rest assured that if you study these pages, their tricks will have no effect on you.

I am surprised that the literati I so admire—Anne Tyler, John le Carré, Margaret Atwood, and the incomparable Peter De Vries (whose poor imitation you might have recognized by now)—have not already mined the rich lode of the deceptive chartmaker's craft as a plot device. Having reeducated myself in these dark arts in the process of writing this book, I now realize that I have a unique artistic opportunity. You see, there was this nefarious chartmaker who alone had the knowledge that the CEO was color-blind—strike that—visually challenged, colorwise....

Gerald E. Jones
Santa Monica, California
August, 1995

The Numbers Don't Lie

Or Do They?

No doubt you've heard a colleague defend a business decision with the sweeping declaration: *The numbers don't lie!* Surely, a conclusion based on solid data *must* be reliable. Now, if you swallow this little bit of wisdom whole, you might also be tempted to believe that charts based on those same numbers must also show the truth, the whole truth, and nothing but the truth.

Oh, if only our world were so simple!

This book can teach you how to lie with charts. Or, if you don't tell lies, you can learn how to defend yourself against the shameless people who do. In this chapter, I'll explain how you or anyone can begin by lying with the numbers—that none-too-solid foundation on which your charts are built.

Have I given you the wrong impression? Please don't think that I want to make it easier for you to be dishonest! Your sense of business ethics is not at issue here.

 Here's the major premise of this little book: Learning how to lie with charts will help you spot deceptive tricks, whether in someone else's charts or in your own. Where you go from there is up to you!

A chart is a graphic presentation of a set of numbers. A mathematician would refer to them as a *data set,* or *data series.* The set of numbers represents a collection of measurements. These measurements typically describe how some *thing*—say, the growth of a company or a wart on your toe—is changing over time. Another type of chart can show the same measurement taken for several things at some "snapshot" in time.

All charts distort the truth, if by *truth* you mean *reality*— the physical world. Charting is a type of *data reduction* whereby the complex world impinging on our senses can be made simpler, and thus easier to understand. Data reduction isn't necessarily a bad thing—perhaps you've heard that *less is more*—but you have to be careful doing it.

The opportunities to distort reality don't begin when you reduce a set of data to a chart, however. There can be all kinds of distortions in the numbers themselves. In fact, the apparently simple act of *counting* can produce serious errors.

Counting Is Simple—Isn't It?

We perceive reality by measuring its effects on our senses. For example, we can count the things we see. However, just the act of counting—without doing any higher math—requires you to interpret, and possibly to distort, reality.

Okay, you walk into the produce section of your local market and start counting pieces of fruit. When your task is done, you hand me a number:

This number is a piece of data. Like garden produce, the data is *raw*. It is *unprocessed*. Raw data is usually, well, downright indigestible. Digesting (understanding) it involves that fearful term —*data processing*!

Now, you can't avoid the processed stuff. You need it just to survive. Face it, reality is messy. It can involve lots and lots of numbers—computers full of them. And handling all those numbers can be inconvenient. They arrive all jumbled, just the way reality generates them—not in nice, neat rows and columns, much less in pretty charts.

By the way, what did you mean by that number you gave me?

19 WHAT?

Are we talking apples or oranges here? Even in the act of collecting the data, you must impose your own notions on reality: Okay, you counted apples. Did you count just the red apples or did you include the yellow ones, too?

Right away, I can see a problem with *categories*. Just trying to count things means we have to make decisions about categories—which things to include and which to exclude from the count. For example, how about that half-rotten apple over there? Surely, you didn't count that. You can't really call it a piece of fruit. After all, it's not a whole piece. And, if nobody wants to eat it, would you call it fruit? Or just garbage?

My Apology to the Citizens of Rome

The word *data* is the plural of the Latin *datum,* meaning *fact.* So it would be perfectly correct to say, "The data are in Harvard Yard." However, even careful lexicographers admit that data can be a collective noun (referring to a group containing more than one datum), in which case it would still be impeccable usage to say, "I went to Georgia Tech, and the data is wherever I say it is." In this book, I follow the crowd in saying "data is," but I will defend not quite to the death your right to say it either way.

Is There Any Truth in Labeling?

The categories of things can be indicated by labels, which are as necessary to reliable data as the numbers, or *values*. As a rule, wherever you see a number, you should find a label nearby.

19 APPLES!

In a chart or in a printed report, if you see a number without a label, stop right there. Demand a label, and don't take "None" for an answer. You have a right to know what you're dealing with, after all. There's more advice about labels in Chapter 10, "Words for Your Charts, Or Is There Any Truth In Labeling?"

My high school chemistry teacher Mr. McNeill was a fanatic about labeling. On exams, if you left a label off an answer, he gave you no credit for it at all, even if you got the number right.

3.6592 ⟵ WHERE'S THE LABEL, YOU KNUCKLEHEAD?

McNeill preached that a number without a label is meaningless. (Well, that's not strictly true. It has a meaning in math, but that was Miss Bowers's class.) By itself, a number is not useful.

Often a label must have two parts—a unit of measure and a description of the thing being measured:

3.6592 LITERS OF $H_2SO_4^*$

*SULFURIC ACID – CAREFUL!

Getting back to your apple-counting, did you count 19 apples or 19 pieces of fruit? Was it 19 pieces, 19 dozen, or 19 cases?

Add a label to a number, and you've really got something:

Leaving a label off an important number is a favorite trick of obfuscators and graphic deceivers. Worse, using a label that is only slightly incorrect is the mark of a truly skilled liar. For example, if I were trying to exaggerate a count, I might say 19 *crates* instead of 19 *boxes*. *Crate* seems bigger, more substantial, than *box*, giving the impression that it can hold more, even though there might be no difference in the physical capacities.

Another set of labels can relate this stuff to the things around it:

TOTAL INVENTORY: 19 CRATES OF APPLES

A naked number is one kind of raw data. It's not much use by itself. Add a label, and the data describes something in the real world. Relate that description to the things around it, and you have useful information. And, after all, getting useful information should be the goal of every honest chartmaker!

What Is Information?

One definition of information is that it answers a question:

Funny about information. Because our brains find it useful, it often stimulates more questions than it answers:

- How many apples did we have yesterday?

- How many apples did we sell today?

- How many apples should we order for tomorrow?

- Who wants this half-rotten apple?

Data or Information?

Here's a familiar example:

DATA	5105238233
INFORMATION	(510) 523-8233

The raw data is a string of digits, which might represent anything. To be useful, the data must be interpreted. The useful form—information—is obviously the second one. Each part of the phone number—the area code, the exchange, and the extension— acts as a label, or data category. The categories are the keys to interpreting the data. (Dial this number and they'll answer, "Sybex"—a publisher with the vision to bring this book into print!)

With the exception of the bothersome rotten-apple issue, one chart might be able to answer all these questions. A single chart might even describe an entire apple inventory at a glance, as shown in Figure 1.1.

Even though the chart addresses all these questions—a real cornucopia of information—its data content is *reduced,* as compared to the raw data that was generated from counting apples. No matter how much you study the chart, you cannot know everything about the real produce bin in the store! For example, you don't know from the chart whether the yellow apples were counted along with the red ones, and you certainly don't know about the characteristics of individual apples. How many of them, for example, are bruised, blemished, or rotten?

To make decisions based on such a chart, you have to make some assumptions about the underlying data. Could I assume that you only counted salable apples? In the end, anyone who didn't do the counting would have to trust the person who made the chart.

Figure 1.1 *This snapshot of the inventory shows how many apples will have to be ordered today to replenish the stock to 36—yesterday's starting level. (Even this simple chart distorts reality, because you can't order apples by the piece. You have to buy them in case lots!)*

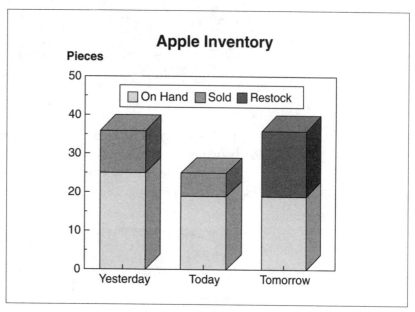

Avoid Passing Summary Judgment!

An everyday kind of data reduction is an arithmetic *summary:*

$$
\begin{array}{r}
19 \text{ APPLES} \\
26 \text{ ORANGES} \\
+ \quad 7 \text{ BANANAS} \\
\hline
52 \text{ PIECES OF FRUIT!}
\end{array}
$$

Who said you can't mix apples and oranges? Summarizing a set of numbers to produce a total—the infamous bottom line—reduces the complexity of reality. In this case, the reduction is down to a single number. The total is more understandable than a collection of raw data. It's tempting to work with totals because we can quickly get

Remember, though, that the process of data reduction obscures the underlying details, which may include important facts. For example, consider the business manager who wants the bottom line on monthly sales. She wants the total, and quickly:

Pleased with this number, the manager might move on to other, more pressing matters. However, as in horror films, danger often lurks just beneath the surface. Here is the raw data summarized by that total—results according to each salesperson on her staff:

THIS MONTH'S SALES
BY SALESPERSON

D. AVALOS	$45,281
P. NOONAN	0
A. ONO	2,919
F. BERG	362
TOTAL	**$48,562**

A deeper inspection of the data reveals that two of the salespeople haven't done much at all, and another had only modest results, while one of them has racked up most of the sales. What's the story here? Are the other three loafing, or have they been developing good prospects that might generate sales next month? Is that stellar performer overworked, perhaps pushing toward a nervous breakdown? The manager would certainly ask these questions if she looked closely at the data, but she could ignore a possibly precarious situation if her focus stays only on the bottom line.

Compare the amount of information you get in the two charts in Figure 1.2. The top chart shows a bar graph of the bottom-line results—total sales. The bottom chart shows the same totals, with the addition of *segmented bars* to show individual contributions. The sales manager would do well to insist on seeing the data presented this way.

The tricks of data reduction go way beyond generating totals. Perhaps you've heard your colleague refer to "massaging" the data? Well, one way to get a good numeric rubdown is to start playing with your chart data as *averages.*

Figure 1.2 *Two views of sales results: The use of segmented bars reveals a hidden story.*

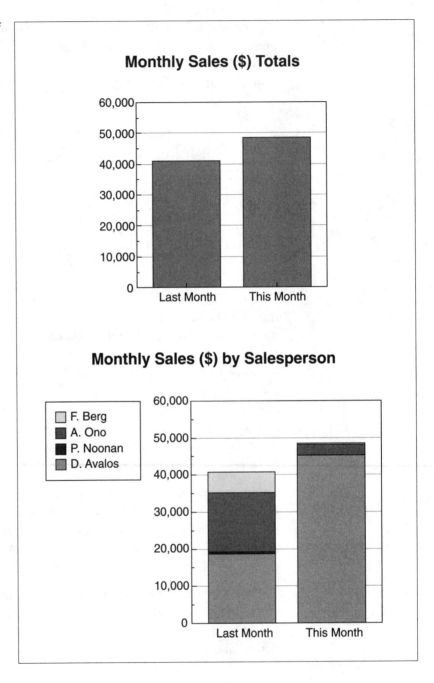

You Can't Trust Your Average Golfer!

A player's proficiency at games such as golf, bowling, and baseball is usually described as an average, another type of data reduction. If a golfer at your office says he shoots in the high 70s, he is *supposed* to be saying that, on a typical day, it takes him less than 80 strokes to knock a little white ball into 18 successive holes on a regulation fairway. The rookie liar, of course, adopts his low score from one exceptional game as his lifetime average. But if you want to lie to the pros, you're going to have to do better than that!

The average is the sum of the scores of previous games, divided by the number of games:

$$\frac{79 + 81 + 78 + 80 + 76 \text{ STROKES}}{5 \text{ GAMES}} = 78.8 \text{ AVERAGE STROKES PER GAME}$$

The average is meant to describe a typical, or usual, score over a player's history. Like summaries, averages sacrifice the underlying detail in favor of a single, more understandable number. And as in the bottom-line sales example, relying exclusively on an average can lead you to incorrect conclusions.

A big problem with using an average as an indicator of performance is that averages are reliable only if the underlying data values are consistent. For example, a golf average of 83 is a fair picture of a golfer who had the following history:

$$\frac{85 + 83 + 82 + 84 + 83 \text{ STROKES}}{5 \text{ GAMES}} = 83.4 \text{ AVERAGE STROKES PER GAME}$$

However, the same score—even though correct as an average—is a poor description of this player's performance:

$$\frac{74 + 73 + 72 + 100 + 98 \ \text{STROKES}}{5 \ \text{GAMES}} = 83.4 \ \text{AVERAGE STROKES PER GAME}$$

This unfortunate player was shooting in the low 70s (a real pro), then slammed a car door on his hand on the way to the fourth game. As a result, his average is *skewed high*, distorted upward by the high scores he shot when he was recovering from his injury.

Similarly, a single average score won't tell you if a player is erratic—perhaps performing brilliantly on rare occasions, dismally the rest of the time.

An average will be more reliable if: 1) there is a large number of individual data values, thus minimizing the effect of any one of them on the result, and 2) the data values are fairly consistent, or without wide variations, thus reducing the effects of unusual values on the result.

There are all kinds of ways of manipulating averages so that they are closer to your own idea of what's typical:

Throw Out the Highs and Lows

Assuming that very high or very low scores are too unusual, just disregard them and average the rest. A pitfall here is that the overall average will be based on fewer games, reducing the reliability of the average. To be fair about comparisons with other players, you could make a rule that you will throw out the same number of games for everybody, even for consistent players.

EXCLUDE HIGH AND LOW SCORES

$$\frac{74 + 73 + \cancel{72} + \cancel{100} + \cancel{98} \ \text{STROKES}}{2 \ \text{GAMES}} = 73.5 \ \text{AVERAGE STROKES PER GAME}$$

Use the Mean

Take the average of only two values, representing a high and a low, by adding them and dividing by 2. The result is the *mean—not* the overall average. The mean might be a better indicator if the data values are erratic.

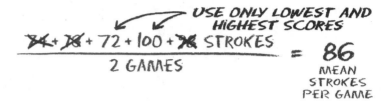

USE ONLY LOWEST AND HIGHEST SCORES

$$\frac{74 + 78 + 72 + 100 + 78 \text{ STROKES}}{2 \text{ GAMES}} = 86$$

MEAN STROKES PER GAME

The mean is sometimes called the *mid-point,* but that's too easy to confuse with *median,* which is something different.

Use the Median

In a group of data values, the *median* is not an average, but one of the values that seems "normal." The median value in a set is located exactly in the middle *of the count,* where there are the same number of values above it as are below it. So, if you have a set of five scores, there will be two scores above the median and two scores below it.

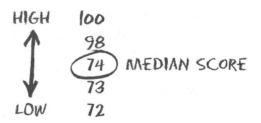

HIGH   100
         98
        (74) MEDIAN SCORE
         73
LOW   72

If you have an even number of data values, you would take the mean of the *middle two* as the median. For example, in a set of four values, you would take the mean of the second and the third values, so that there are two values above the mean and two below it.

Figure 1.3 *The smoother line is the player's four-game moving average, which some people might regard as a better description of overall progress.*

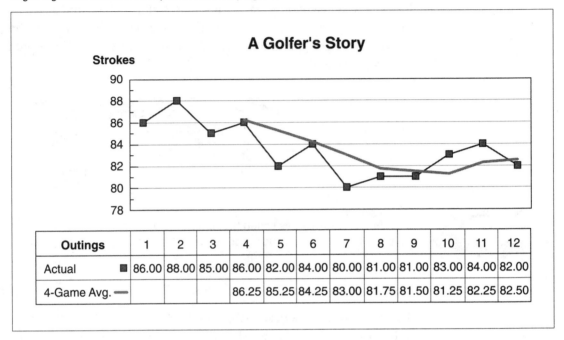

Outings	1	2	3	4	5	6	7	8	9	10	11	12
Actual ■	86.00	88.00	85.00	86.00	82.00	84.00	80.00	81.00	81.00	83.00	84.00	82.00
4-Game Avg. —				86.25	85.25	84.25	83.00	81.75	81.50	81.25	82.25	82.50

Use a Moving Average

Averaging all your golf scores over your lifetime would be unfair. You don't want your recent scores to be distorted by your poor beginning scores, which should become less and less typical as you improve your game. Instead, a *moving average* might include only your most recent set of games—say, the last four. As each new game is included in your average, the oldest game drops off so that there are always just four games included. A chart of your lifetime performance, then, would show not only your actual scores, but also a series of four-game averages (see Figure 1.3).

Don't Count Blanks

Whenever you are working with averages, you can get into real trouble if there are any blanks, or gaps, in the series of data values. In the case

Will Those Lying Golfers Stop at Nothing?

If playing with their averages doesn't yield the right numbers, golfers have yet another way of manipulating their scores—the *handicap.* For a less-experienced player, a handicap is some number of strokes that always gets subtracted from that player's actual score. The golfers rationalize it this way: The score adjusted by the handicap compensates for widely different levels of skill so that it is possible to make comparisons with scores of more experienced golfers. So, here's an important lesson in business math: When none of the conventional methods produces the answer you want, *invent a new method!* (Be sure to give it a clever name that invites sympathy.)

of golf scores for several players, a blank might represent a day when one of the players didn't show up. However, a position that holds a zero is not a blank. In general, it is good practice to use zeros rather than blanks in a data set. When counting how many values will be averaged (the number you use as a divisor), *exclude the blanks* but *include any zeros.* Remember that zero values will distort the average unless the other values are near zero (0.1, say).

These are just a few of the many ways you can fudge—er, manipulate—data to reduce a set of values to one value that's typical, normal, simple, digestible, understandable—but nonetheless misleading. In the last few examples, I used various methods to get typical golf scores of 83.4, 73.5, 86, and 74 *for the same player!* Each of these techniques is a perfectly valid type of data reduction commonly used in developing data for business charts. The particular technique employed is mostly a matter of judgment.

Which score do *you* think should be regarded as typical for the player with the injured hand?

LIAR'S TRICK   Beware of obfuscators who use the term *average* loosely to refer to any of these very different techniques!

Trendy Thinking Can Be Risky

Just as dangerous as data reduction is any attempt to *expand* a set of data, either by guessing in-between values *(interpolation)* or by identifying trends and projecting future values *(extrapolation)*.

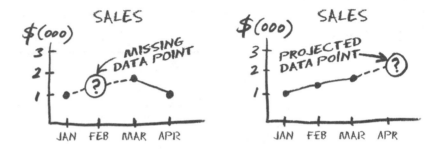

Using charts to identify trends is probably the slickest of the expert liar's tricks, and this stuff can get ambitious, mathwise. I spill my guts on this subject in Chapter 8, "Trends—Or, A Conservative's Guide to Trendy Thinking."

The Bottom Line on Data Reduction

I admit that this chapter offers many problems, with few solutions. But I want you to develop a healthy sense of caution about drawing conclusions from charts and chart data before we get down to the details in the chapters that follow. My intention is to begin by making you sniff the air, so to speak, for those rotten apples.

As examples in the rest of this book should demonstrate, charts aren't simply neutral vehicles of communication. They don't just show data. Charts are ways of reducing *and interpreting* data to accompany a business message.

You will learn that there is at least one interpretation of the data in any chart, *even if its designer did not put it there deliberately.*

The trick, if you'd call it that, is to make a chart's interpretation of the data *support and reinforce* your business message. That's not necessarily lying. It's just good communication.

Pies

Or, When in Doubt, Throw One at Them!

Everyone wants a piece of the pie. The *pie chart* is easy to make, easy to understand. Right?

The familiar pie chart is the most overused, misused, and sometimes downright useless trick in the presenter's repertoire. Liars love them, and in this chapter I'll show you why. You'll also find out how to throw one at your audience—if you really must—without making an awful mess.

The "Why" of Pies

This type of chart is, yes, easy as pie—a circle cut into slices:

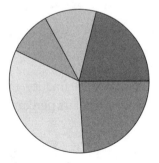

 Other generic terms for the slices of a pie are *wedges, segments, cuts,* or *pieces.* The correct term in geometry for the slices of a circle, however, is *sectors.* (Okay, who cares?)

There's a simple rule for using pie charts correctly:

Instead of percentages, you can use a pie chart to show any proportional relationship between a slice and the whole pie. Such a relationship can also be expressed as a fraction, a ratio, or a decimal. For example, "This piece is one-fifth of the pie"

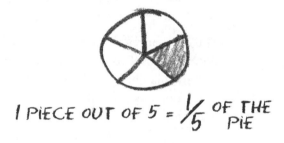

can be written

AS A FRACTION: ⅕
AS A RATIO: 1:5
AS A DECIMAL: 0.2

Multiply the fraction by 100—the number of percent in a whole pie—and you get its percentage of the whole:

$$\frac{1 \text{ PIECE}}{5 \text{ PIECES}} \times 100\% = 20\% \text{ OF PIE}$$
(WHOLE PIE)

Showing any of these relationships—percentages, fractions, ratios, or decimals—is the purpose of pies. But you will be inviting liars to your table if you serve up your pies as anything else!

The trouble starts when you begin to think of pies in terms of *amounts*. A symptom that you've slipped into this mode of thinking is showing amounts in the slice labels. But, you ask, what's wrong with providing such potentially helpful information?

Percent *of What?*

In the first chapter, I strongly advise you to put labels on numbers. A number without a label, so I say, is useless. Now, pie slices should have labels, too. You and the audience have a right to know what you're dealing with.

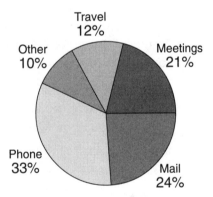

Okay, fine. It's clear what each of the slices here represents—some percentage for each labeled slice. But, you all want to know,

- What does the whole pie represent?
- What is *its* label?
- These slices are percentages *of what?*

Figure 2.1 *The label for the whole pie should be the title of the chart.*

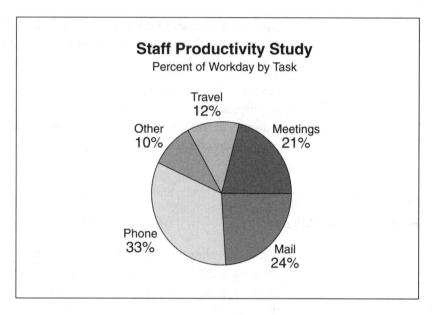

The label for the whole pie should be the title of the chart, as shown in Figure 2.1. In this example, each slice represents some portion of the total hours in a workday. The chart, then, is meant to show how the members of a sales staff are dividing their time among required tasks.

The whole pie represents the amount of time in a person's workday. So, maybe you now want to know, "How many hours are there in a workday?" Or, "How much is the total pie?"

In a pie chart, the *amount* of the whole is hidden from you! This secret feature of the pie chart can be a dirty little trick —or a redeeming virtue, depending on how you use it.

The purpose of a pie chart is to focus the attention of the audience on percentages—on the *relative* sizes of the slices. The actual values of the slices and of the whole pie should not be important. The answer to "percentage *of what*?" should always be "of the whole pie." If you need to know more about the *value* of the whole, *you shouldn't be using a pie at all!*

Look back at the time study in Figure 2.1. Why can't you ask, "How many hours in a workday?" You might assume that a workday is eight hours long, but, in this case, you shouldn't. This chart is just as valid for the part-time staff who work just two hours as it is for the overachievers who put in twelve!

The intent of this chart is to support a discussion about how much time salespeople spend traveling to meetings in relation to all their available time—the whole pie—and in relation to the meetings themselves. The total amount of time—the number of hours each day—is not important to the intended message.

What is wrong, then, with the chart in Figure 2.2?

The pie charts in Figures 2.1 and 2.2 are identical, except for the slice labels. The first chart shows each slice, or type of task, as a percentage of a total workday. The second shows each slice as an *average number of hours* in a workday.

Figure 2.2 *This pie chart shows values instead of percentages for the pie slices. It gives the possibly incorrect impression that most people on the staff work 7.5-hour days.*

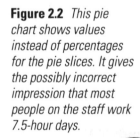

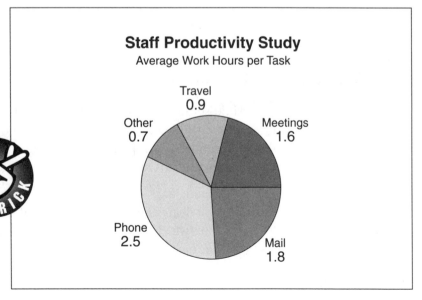

Staff Productivity Study
Average Work Hours per Task

Travel
0.9

Other
0.7

Meetings
1.6

Phone
2.5

Mail
1.8

When actual values are given for pie slices, the audience will be tempted to add them to find out how much the total is. (If they don't do this, they will at least have some mental impression of the total, even if they are not fully aware of it.) In the case of the time study, the total number of hours is 7.5—not 8. Again, these are average hours. The part-time people have been included with the overachievers and everybody else. In fact, the majority of the staff works full 8-hour days. The length of the workday in the chart has been skewed by including the others in the average.

Look back at Chapter 1 to see how deceptive averages can be!

The presenter's intention is to recommend to top management that salespeople try spending more time on the phone and in electronic conferencing and less time traveling and in face-to-face meetings. However, the presenter could get easily sidetracked by irrelevant questions if the top managers become more worried about how the staff is spending the other half-hour each day—the 0.5 hour that seems unaccounted for in the daily average of 7.5!

Resist the temptation to put values instead of percentages on the slices of a pie chart. If the actual values are essential to your business message, use another type of chart.

Pies, Clocks, and Compasses

A possible source of confusion about pie charts has to do with the *orientation*, or positioning, of the slices. There are two schools of thought on how to plot and read a pie chart. I'll call the two schools the "mathematicians" and the "clock watchers."

The mathematicians, following the rules of geometry, divide a circle (pie) into 360 degrees. The starting point of the circle is at 0 degrees (same as 360), which if you were looking at the points of a compass, would be due east:

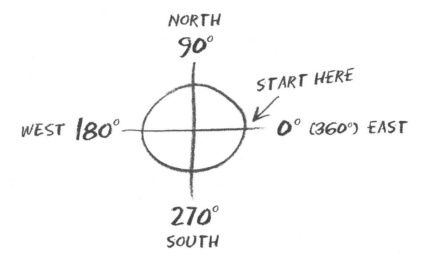

When they plot a series of percentages as pie slices, the mathematicians start at 0 degrees and proceed *counterclockwise*:

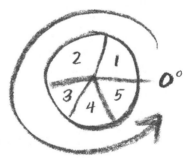

Some people, perhaps the mathematicians and their ilk, have been trained to read pie charts this way—counterclockwise. The clock watchers, on the other hand, think of a pie as a clock face:

The starting point is not at the side, but at the top—12 o'clock. Slices are plotted around the dial—you guessed it—in *clockwise* order. This is also the order in which the clock watchers read the data from the chart.

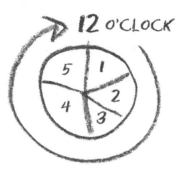

Despite the fact that more people know how to tell time than know geometry, most computer graphics programs use the mathematicians' method: *Start at 0 degrees and go counterclockwise.* Your audience, however, will most likely be populated with clock watchers: *Start at 12 o'clock and go clockwise.* Be aware of any automatic decisions the graphics program may be making for you, and adjust the results accordingly. Make your pies the way the important people in your life like them!

Which Slice Is Most Important?

Does the way a pie is built affect how the audience reads it? After all, don't people see the whole pie at a glance? They don't necessarily read the slices in any order. You might also suspect that the eye will be drawn to the biggest slice. The importance your audience gives to a particular slice, however, can depend on its *position* in the pie, regardless of its size.

Many people will assume that the most important slice is on the right side of the pie. To the mathematicians, that's 0 degrees. To the clock watchers, it's 3 o'clock:

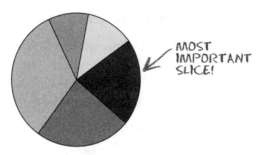

The importance of this slice can be emphasized further by *exploding* it, or pulling it out from the rest of the pie:

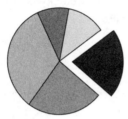

 Whether you are a clock watcher or a mathematician, the most important slice is always the one in the upper-right-hand position, *with this exception:* Regardless of its position in the pie, *any* piece that gets exploded becomes the most important piece!

A friend of mine made the following observation about the relative importance of pie slices:

> I was reading the *New York Times* this morning and discovered for myself how the 0 to 90 degrees position is where the eye seems to be drawn first. A pie chart showed the racial breakdown for people on welfare, and my eye was drawn to the group labeled "White" on the right-hand side, even though the percentage was about the same as the group labeled "Black." (The two groups combined are about 76 percent of the total.) It occurred to me that if the African-American group had been put instead in the top-right position, the chart might have emphasized a different point. My reaction—with the "White" group on the right-hand side—was: "That's interesting. The stereotype is that it's mostly minorities on welfare, but the percentage of white people on welfare is just as high."

The *Times*, by the way, did not fall into the trap of showing numbers— they showed only the percentages.

What Determines the Order of Slices?

The order of pie slices normally follows the order of the values in its series of data points. The order is simply the sequence of the data in list form, as the numbers might appear in a table or spreadsheet. For example, here is a series of percentages:

Slice	Percent
A	20
B	40
C	15
D	25

Using the counterclockwise mathematicians' method, this series will be plotted as:

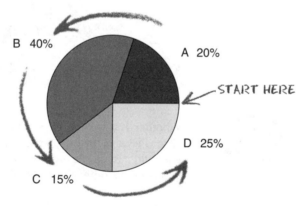

Still proceeding counterclockwise, you might prefer to plot them in order of size:

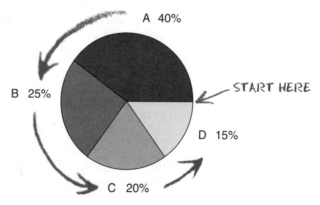

Assuming that the 40 percent slice is most important, it can be emphasized by positioning it at 0 degrees—*rotating* the pie so that the right end of the pie is in the middle rather than at the beginning of the slice. Exploding the slice enhances the effect:

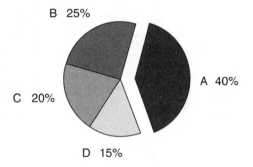

Of these examples, there is no "best" way to draw pie charts. You simply need to be aware of the hidden assumptions behind their construction. For example, if I were trying to downplay the amount my company was spending on vacations for the senior executives and play up the amount spent on charitable contributions, I could place the Charity slice in the right-hand position, exploded. Charity would then be emphasized, even if its percentage were smaller than the Vacations slice.

Which Slice Is Most Important? (Revised Version)

Now, the foregoing advice is all sound—until you start dealing with so-called *three-dimensional (3D)* pies. A 3D pie appears tilted on its edge so that it has thickness. The pie itself is no longer a circle—it's distorted, squashed into an ellipse:

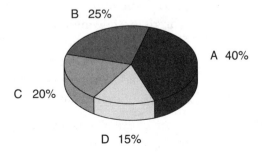

Figure 2.3 *This 3D pie really distorts the importance of the bottom slice. The thick edge makes the Meetings slice look much more substantial than the Phone slice, even though its percentage is smaller.*

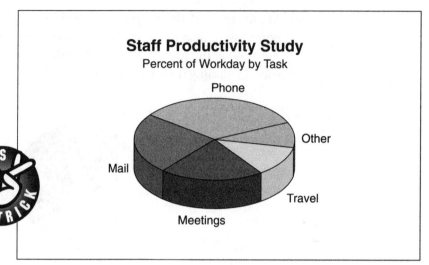

In a 3D pie, people will usually think that the *bottom slice* is the most important. That's because the dimensional effect distorts the apparent size of the slice by literally giving it an edge. People will perceive this slice as bigger than it would appear in a circular, two-dimensional pie.

Here's one of the liar's favorite tricks: Put the slice you want to emphasize at the bottom of a 3D pie. Give the pie some real thickness and tilt it way back. And, to compound the crime, omit all data labels. The effect is awesome, as you can see in Figure 2.3.

Not Using Pies to Show Sales Results

Perhaps the most common—and the most offensive—misuse of pie charts is for showing sales results (see Figure 2.4). This example violates my cherished rule about pies: The slices are labeled as dollar values rather than as percentages. The audience will be strongly tempted to mentally add the slice values to find out the total sales dollars, especially if the amount has not been shown previously in the presentation.

A correct use of a pie would be to compare market share percentages among competitors, as shown in Figure 2.5. This is the same pie as

Figure 2.4 *Including dollar amounts in pies that show sales results is usually a bad idea because it distracts from the real story—the proportions of the slices to the whole.*

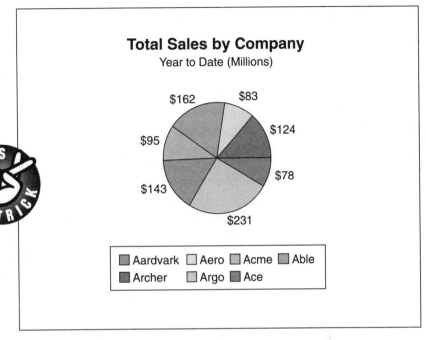

Total Sales by Company
Year to Date (Millions)

Figure 2.5 *Here is the chart from Figure 2.4, but with percentage labels instead of dollar amounts. When pie slices are labeled with percentages, the audience will focus correctly on the relative importance of the slices rather than on their explicit values.*

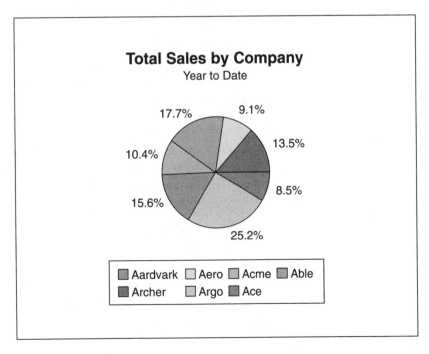

Total Sales by Company
Year to Date

Figure 2.4, but with percentage labels. The audience is not tempted to ask about the total sales dollars—the volume is not the issue. Instead, attention is focused on the percentages of the total market.

Don't Tell the Two-Pie Story

Another trap opens wide for you when you proudly carry your pies into the world of sales. Whenever you are dealing with sales results, you can't expect your audience to remain focused on the present. They also want to know about the *past* and the *future!* Today's sales results are always measured against yesterday's—or last month's or last year's. And, before you finish your presentation, the audience will naturally be interested in your projections about tomorrow's sales.

Sales results always get this yesterday-today-and-tomorrow perspective. That's because sales dollars represent the ongoing income of the business—its very life. *Everyone* in a business is interested in actual sales dollars because fluctuations in income affect everyone's financial well-being.

For these reasons, one pie used to show sales dollars can quickly become *two*—LAST YEAR and THIS YEAR—on the same chart, as shown in Figure 2.6. Bolder presenters will venture to try *three*—including another pie with estimates for NEXT YEAR.

This two-pie comparison is confusing because it uses sales dollars—actual values—as slice labels. It breaks the rule about using pies to show only percentages. As with the single-pie example, this chart would make more sense if it were intended to show market shares only —dropping the dollars in favor of percentages.

Nevertheless, heedless tellers of two-pie stories will stop at nothing! Making both pies the *same size* looks wrong, they think. After all, this year's sales are greater than last year's—so, they reason, this year's pie should be bigger! Compounding the error of showing the sales dollars instead of the market-share percentages, they adjust the sizes of the pies, as shown in Figure 2.7.

Figure 2.6 *The Two-Pie Story is told by information gluttons who can't resist the temptation to have that second pie!*

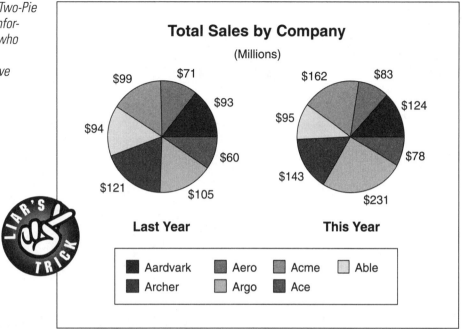

Figure 2.7 *Using different-size pies just makes a two-pie chart that much more confusing.*

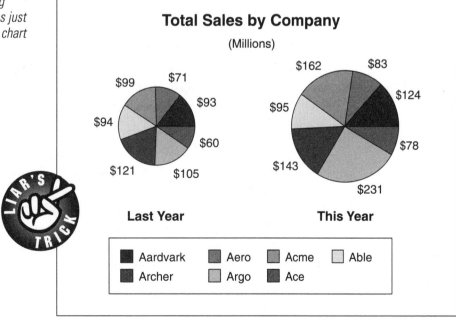

Figure 2.8 *An accurate comparison of the sizes of objects—whether circles or squares—is by area, not by height.*

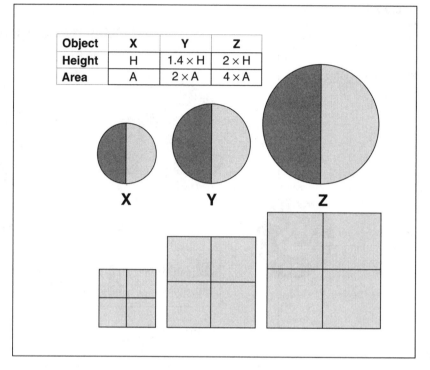

Object	X	Y	Z
Height	H	1.4 × H	2 × H
Area	A	2 × A	4 × A

Some people will be perfectly happy with the chart in Figure 2.7. It's a feast for fools, though. Here's why.

If this year's sales are 1.4 times greater than last year's, the chart designer's inclination will be to treat them like bars and make the height (the *diameter)* of the This Year pie 1.4 times that of the Last Year pie, as shown in Figure 2.7.

However, an accurate comparison of the sizes of objects—whether circles or squares—is by *area,* not by height. In Figure 2.8, the area of the pie on the right is actually *twice* that of the pie on the left. The correct sizing of the second pie—which is a smaller area—is shown in the center. The two-pie chart is misleading because the second pie is larger than it should be, inflating the sales dollars it represents. The true relationship of the areas is easier to see when they are drawn as squares, as shown in the lower part of the figure.

Further Secrets of Pie Thieves

Liars and pie thieves have a few other tricks that they can do with Mom's apple pie:

Adjust Overall Pie Size

Even if you are showing just one pie, you can distort the importance of the whole issue by playing with the size of the pie. Liars who want to diminish whatever the pie represents will often clutter the chart with lots of labels, forcing the pie to be smaller to fit on the page or screen.

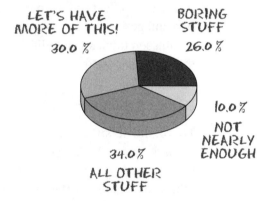

Abuse the All Others Category

In a pie chart, you are forced to show the whole thing, whatever the thing under discussion is: The whole pie always represents 100 percent. Liars sometimes find it inconvenient that all pieces of the pie must be accounted for. They typically get around this by labeling a mystery slice All Others.

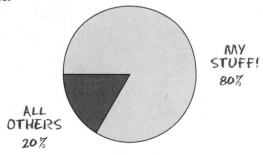

Watch out when you see All Others, especially if the slice is more than 10 or 15 percent of the whole pie. The presenter just might be hiding a big one in there, rather like burying your vacation in the Business Travel category of your expense report.

There can be legitimate reasons to have an All Others category. For example, if you are showing market share, there could be a dozen competitors that share just 10 percent of the market—and, for the moment, none of them is important to you.

Don't Show the "Don't Knows"

If the All Others slice complicates your story, simply exclude it from the pie! The effect will be to increase the sizes of the other slices. Their percentages will be increased because you are reducing the amount of the unseen total. Most audiences won't catch on that you're not telling them the *whole* story!

Fudge Small Slices

There is one situation in which I *recommend* that you cheat in preparing a pie chart. If you must include a very small slice—equivalent to 1 percent or less—I recommend plotting it at about 1.5 percent (about 5 degrees of the circle) so that the thin slice can be seen. Otherwise, a thin slice looks too much like the *callout lines* that some designers use to connect labels to slices. You should include a helpful label nearby that clearly shows the true percentage. Exploding the slice, as shown here, may help you make your point.

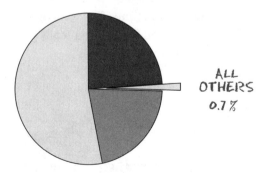

Exotic Pies for the Info Gourmet

Two variations of the pie chart can be especially tempting on special occasions: the pie-column chart and the donut chart.

The Pie-Column Chart

This hybrid chart format combines an exploded pie with a *column*. The column is a type of stacked bar chart, its segments stacked on top of one another. The column presents an itemization of the exploded slice, as shown in Figure 2.9. This example is a solution to the All Others problem described above: Show the market shares of all those tiny competitors in a column.

You might also use a column to itemize and focus in on a breakdown of the biggest or most important pie slice.

Some people use a second pie instead of a column, but I think the columnar design is less confusing to the audience. It will be obvious that the column is not a second version of the pie.

Figure 2.9 *In a pie-column chart, the column itemizes the exploded slice.*

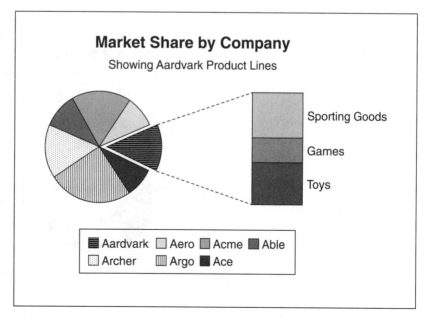

The Donut Chart

Some clever designer decided to hollow out a bunch of pies and show them stacked inside each other—like rings on an archery target. The prize-winning name for this concoction is the *donut chart,* as shown in Figure 2.10.

Now, I'm not sure what use a donut chart might have for you, except perhaps to perplex your audience. (Liars take note!) It's just that pies are so simple and basic—easy to digest, you might say—why not use a series of pies on different charts rather then trying to cram them all around the same little donut hole?

But, if your audience likes these donuts, go ahead and feed them what they want. Your presumably legitimate purpose might be to compare the percentages, or ratios, of several different entities. For example, look at Figure 2.10. If you were showing market shares, each ring on the donut might represent comparative market penetration in a different product line. Just remember to give your audience plenty of time to study this one! And, as with pies, avoid labeling donut slices with explicit values.

Figure 2.10 *A donut chart is a set of concentric, hollowed-out pies.*

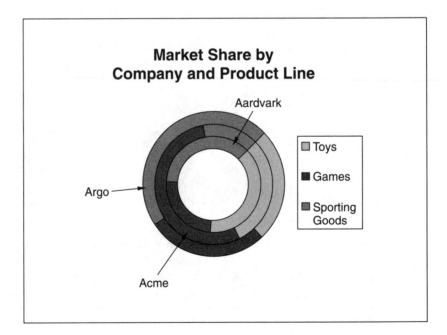

Honest Alternatives to Rotten Pies

Remember, pies are for percentages or ratios. If you find yourself putting actual numbers (such as dollars) on a pie, you might be able to clean up your act by using bars or areas instead. Here are some of the alternatives:

Stacked Bars

As shown in Figure 2.11, this is the best way to show relative sizes of different segments as well as their actual amounts. (There's more about these stacked things in Chapter 4.)

100 Percent Bars

In this format, all stacked bars are the same height, each representing 100 percent (see Figure 2.12). Think of them as pie dough rolled into

Figure 2.11 *The stacked bar chart is a good alternative to showing pies with actual data.*

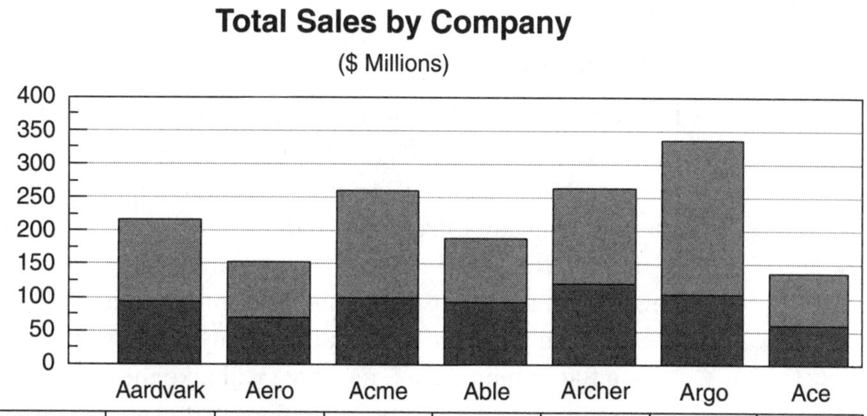

Total Sales by Company
(\$ Millions)

	Aardvark	Aero	Acme	Able	Archer	Argo	Ace
Last Year ■	93	71	99	94	121	105	60
This Year ■	124	83	162	95	143	231	78

Figure 2.12 *The 100 percent bar chart can be used instead of a donut chart or instead of multiple pies.*

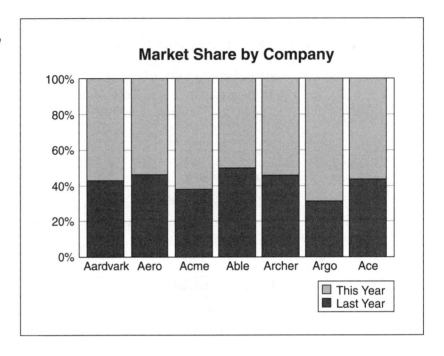

same-sized bars. As with donut charts, you can compare the equivalent of several pies in one chart. As with pies, the labels should be percentages, not actual values.

Stacked areas

Like stacked bars, this format can be used to show both relative sizes and actual values (see Figure 2.13). Deciding the relative positions of the areas can be confusing to an audience, however. Are the areas stacked vertically—like bricks and mortar—or are they layered on top of one another—like drawings on separate transparent sheets, each new area starting on the same horizontal line at the bottom of the chart? If the areas are truly stacked, only the bottom data series will have a flat baseline. Each new area will take the top of the preceding one as its baseline, and any fluctuations in that line will distort the highs and lows in the series above. (These distortions can be minimized by putting the "least bumpy" areas on the bottom.)

Figure 2.13 *Area charts are alternatives for pies, but where's the baseline of each series? Is it the bottom of the chart, or is it the top line of the preceding series?*

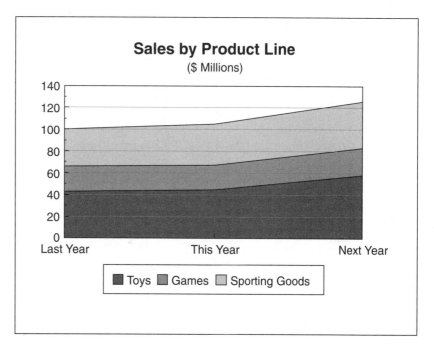

 There's much more about bars and areas—including the stacked kind— in Chapter 4.

Okay, What Are Pies For?

The pop quiz for this chapter has just one question. And you know the answer:

Pies are for percentages!

If by now you're sick of my harping on this point, then I guess I've done my job!

Orientation

Or, What Goes Up Must Go North

CHARTS are pictures. In this chapter, I encourage you to think about them that way—as collections of forms rather than diagrams of data. In particular, I want you to think about *orientation*—the position and direction of graphic shapes in relation to a page or screen. Artists call it *composition,* and it can have powerful effects on an audience, regardless of the data content of your charts.

When you set out to make a chart, you should be aware of some basic assumptions that are shared between the presenter and the audience. These (usually subconscious) assumptions include the meaning and importance of directions in space. For any kind of graphic, designers must make decisions—conscious or not—about how the shapes that compose the image are oriented. Viewers get very different impressions depending on whether shapes appear to go left, right, up, or down.

Where Does the Time Go?

People in Western cultures read from left to right. For these readers, rightward motion—the way the eye scans a page—is associated with the passage of time, and thus with positive movement, or even the idea of *progress.*

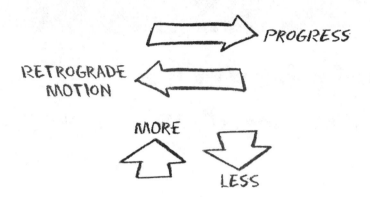

Left Is Just, Well, Gauche

Conversely, movement from right to left is considered backward and negative—*bad!* From these same notions came the root of the English word *sinister*, which means both *left* and *unlucky*.

As you will see, charts that show the progress of time rely on the notion that it flows from left to right.

Up Is More, Which Eventually Comes Down, Becoming Less

In most cultures, movement upwards is associated with increase or gain, downward with decrease or loss. Where charts are concerned, up-down orientation has another effect: People who are used to getting their information from reading text will often begin at the top of a screen and scan downward, even if they are looking at a graph rather than reading words on a page. This downward movement of the eye is not associated with negativity—you don't feel worse about the words at the bottom of the page. It's simply that readers are trained to look from top to bottom, so they will do this just as surely with a chart as with text.

Charts that show fluctuations in amounts rely on the notion that *up* means *gain* and *down* means *loss*.

Concepts of orientation aren't necessarily the same when you are looking at photographs. Scientific studies have shown that eye movements involved in looking at photographs or live scenes are very rapid and complex—especially when the viewer is studying another human face. More about faces at the conclusion of this chapter.

Be Glad Your Audience Is Prejudiced

Much effort in the graphic design field has been devoted to finding more universal means of expression to minimize the effects of biases, which are primarily cultural. Cultural biases are things you learn—rules of a particular society—not instincts you're born with.

International signs constructed of simple, generic symbols are examples of designs that attempt to minimize some cultural biases. The designs aim for simplicity, reducing the details in symbols to the bare essentials—removing from human figures, for example, such details as skin color and apparel, as well as eliminating explanatory text wherever possible:

Because international signs must be cross-cultural, they can't take advantage of orientation. Nevertheless, the effects of orientation can't always be avoided. These biases exist apart from the literal interpretation of a sign. For example, this sign is cross-cultural—even people with different cultural backgrounds will interpret this sign to mean something like: "This way to the exit."

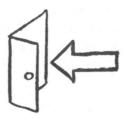

Beyond this literal interpretation, viewers of a sign may respond emotionally to it, according to the orientation of its symbols—in this case, the direction the human figure is walking and the position of the arrow.

It's What You Don't Say

Which of these signs might make a Western traveler feel more "positive" about a choice of routes?

Never mind that there is only one spot to place the sign in an airport terminal and the arrow *must* go the other way. Certainly, you can't go pointing people in the wrong direction just to make them feel better. And no one would ignore a helpful left-pointing sign because of such a vague feeling. But, be aware that people carry those biases with them, no matter which way they decide to turn. You might not have any choice about which design to use in the airport situation, but you could give careful consideration to how you would use the same sign inside a store to direct shoppers to a merchandise display.

When your audience shares a set of biases, a chart will be more effective if its design plays to those assumptions deliberately. For example, you can probably assume that Westerners will interpret rightward motion as progress. *Use that prejudice to your advantage!* Don't try to neutralize their biases, which probably isn't possible anyway. Simply make your designs consistent with what they expect.

It might seem somehow unethical to reinforce these prejudices, but realize that you can't escape them. Graphic conventions are as much a part of language as speech. Your choice of the language itself depends upon many conventions, including these notions of direction. Different sets of assumptions *and visual biases* will apply, for example, to native speakers of Hebrew and Farsi (who read right to left) or Chinese (who read top to bottom).

In short, your charts aren't international signs. You can't make them cross-cultural, and you shouldn't try. They will have more impact if you are aware of cultural biases and play shamelessly to them!

Will You Get Rave Reviews in Tel Aviv?

What should these differences in culture mean to you, in practical terms? For one thing, don't assume that if a presentation made a big hit in the New York office, it will make the same favorable impression at an international conference in Tel Aviv.

Wait, you say, isn't charting an international language of its own? Like mathematics?

Well, chart formats generally transcend language.

But, like the interpretation of international signs, the concern remains how the audience will *feel* about a chart, beyond the literal interpretation of its data.

Why do you need to know these things? Well, if you make a chart that shows rising sales, you want its composition to convey the reinforcing feelings of positive motion and gain. But if you ignore the effects of orientation, your chart might instead convey the opposite feelings of retrograde motion and loss! Ambitious liars can learn lessons here, as well. They can make bad news look not so bad (to an unsuspecting audience).

XY Means Quantity Versus Time (Usually)

A set of charting conventions that play directly to Western cultural bias is the *xy* chart, perhaps the most common format used in business presentations (see Figure 3.1). *The* xy *chart is a powerful device precisely because it is rooted in cultural biases about the meanings of left-right, up-down.* The how-to of charting in *x* and *y* is covered in Chapter 4. My focus here is on the messages that might be hidden in these charts.

Figure 3.1 *To most
Western audiences,
an* xy *chart shows a
relationship between a
variable quantity and the
progress of time.*

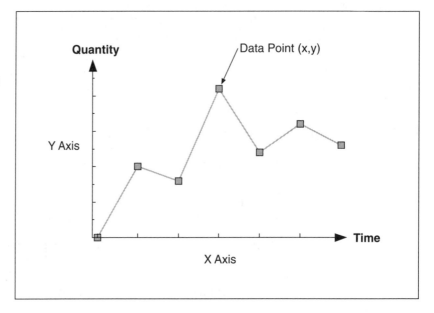

Bar, line, area, and scatter charts are all based on *xy* plotting conventions. They all show *quantity* in relation to *time*. Quantity goes up and down on the vertical axis—the *y axis*. Time progresses from left to right along the horizontal axis—the *x axis*. People who share the assumptions that *up* means gain and *right* means progress (or just "in the future") can learn much about a trend line even if the graph has no precise detail in the form of scales or labels.

For example, the crude graphic in Figure 3.2 conveys a simple message: "The U.S. national debt has been growing especially rapidly in recent years, and threatens to get out of control." You could arrive at this interpretation even if you'd been asleep for the last 20 years and didn't already know this basic fact.

Look again at the text of the message. That's a lot of words, although far short of the thousand the picture is supposedly worth. Still, the text of the interpretation is a surprisingly specific conclusion from little more than a squiggly line and a couple of symbols!

Figure 3.2 *This crude chart, which has no numeric scales or labels to help you interpret it, has a powerful implied message.*

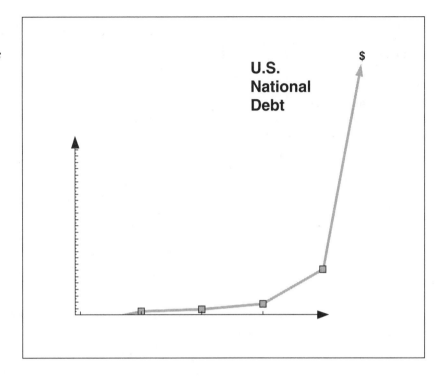

It's Obvious Because It's ... *Um* ... Obvious

The fact that this example seems so obvious is, in itself, proof of the unspoken assumptions it uses so effectively. Stop a moment to consider what you *don't* see in that graph. How do you know it's an *xy* graph? Furthermore, how do you know it's a *vertical* xy (quantity goes up-down not left-right)? Is there anything *in the graph itself* that tells you the events plotted at the left end of the line are further in the past than events at the right end?

Now, look at the chart in Figure 3.3. What is the message *here?*

The answer is: The same as the first chart (Figure 3.2)! They are both plotted from the same set of data.

In the second example, the scales have been rotated so that the plotting orientation is horizontal rather then vertical. Nevertheless,

Figure 3.3 *This chart plots the same set of data shown in Figure 3.2, but because of its unconventional orientation, the intended message is not as obvious.*

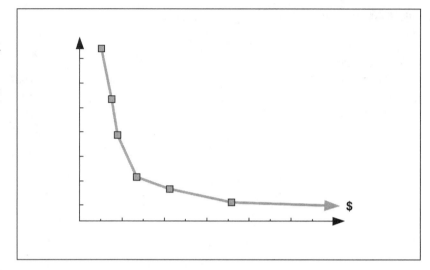

this odd-looking version is a perfectly valid chart. It's just not the format you're used to seeing *for this type of message*—quantity versus time. Perhaps it will become clearer if the odd-looking charts are shown with numeric scales and descriptive labels, as in Figure 3.4.

Figure 3.4 *Adding scales and descriptive labels makes it clear that this chart, despite its odd format, tells a familiar story.*

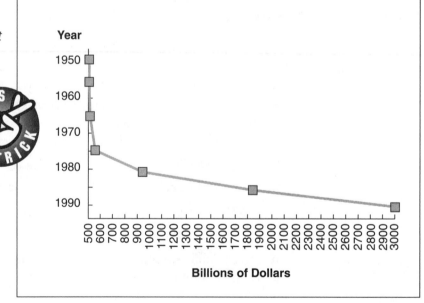

Again, the chart with horizontal orientation is perfectly acceptable—mathematically. But it conveys the intended message poorly or not at all because its orientation defies your common-sense ideas about the meaning of up-down and left-right.

A number of computer software applications are available for producing business charts. Typically, these programs offer a wide variety of charting formats and styles—many of which do not adhere to the time-magnitude relationship just described. A particularly weird format is the *radar chart,* which is so intriguing I cover it separately in Chapter 5.

As a presenter, you depart from convention at some risk. It generally would not be appropriate, for example, to use a horizontal bar chart to represent sales volumes, as in Figure 3.5.

Your intention might be to introduce variety to the presentation, but the result will be added confusion. Horizontal bars are an excellent way to show time spans and durations, however. Figure 3.6, which shows how many hours per week staff spend on various tasks, is a good example of a chart that uses horizontal orientation effectively.

Figure 3.5 *Using a horizontal orientation to show sales volumes fails to capitalize on the intuitive notion that* up *means* more *in a vertical chart.*

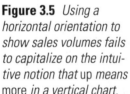

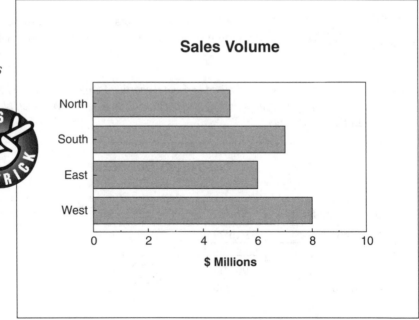

Figure 3.6 *Because
left-to-right motion
generally implies the
passage of time, a chart
with horizontal
orientation is ideal for
plotting time spans.*

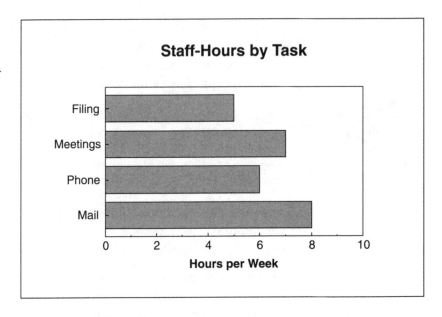

The *Gantt chart,* or schedule chart, is a specific format used in project management for showing task durations and timing relationships as horizontal bars (see Figure 3.7).

Specialized software products have been developed for project management which typically can generate Gantt charts from a

Figure 3.7 *The* Gantt chart *is a specialized horizontal charting format used to show the durations of individual tasks within a larger project. It also capitalizes on the intuitive notion of Western readers that time flows from left to right.*

Project Schedule

ID	Task Name	Start	Mar	Apr	May	Jun	Jul	Aug	Sep	Oct	Nov	Dec	Jan	Feb	Mar
			|		1999										
1	Plan	03/15/99													
2	Develop	03/22/99													
3	Test	04/05/99													
4	Implement	04/12/99													

scheduling process called *time analysis.* The chart in the figure was generated by Microsoft Project. Even though Gantt charts might not be specifically supported by some graphics programs, there is usually something like a *stacked horizontal bar* chart which can be adapted for this purpose. There's more about Gantt charts in the next chapter.

Tempted to Try a New Slant?

So much for right angles. What about slanting lines? What impact do they have?

Any direction on a page or screen can be described in terms of left-right (horizontal position) and up-down (vertical position). *The impact of a slanting line is simply the net effect of its left-right and up-down components.*

For example, you know by now the hidden message in a right-pointing arrow, and you know what an upward arrow can mean. A straight, right-pointing arrow has no vertical component, no magnitude. Therefore, it implies progress in time only. An upward-pointing arrow shows an increase in magnitude, but because it has no horizontal component, or no time span, it implies a static situation. A third design uses an arrow pointing upward and to the right to convey a sense of *both* forward motion and growth, or gain:

This combination of both positive biases makes the slanting arrow by far the strongest visual statement of the three designs. The two positive components *up* and *right* combine to make a doubly positive impression. Likewise, combining two negative components—*down* and *left*—results in a forcefully negative design:

Notions of North, South, East, and West

Linguistic and graphic conventions about left-right and up-down carry over to concepts about geography. These concepts should also be considered in your graphic designs—even in charts that have nothing to do with maps!

From early childhood, North Americans have been taught to visualize the earth in terms of a Mercator projection map with the continental United States at or near the center (Figure 3.8).

Since childhood, Westerners are taught their geography with reference to Mercator. So I learned that North is up, South is down, East is right, and West is left. But there's no objective reason to view the world that way: we're on a globe floating in space, where *up* has no objective meaning at all!

Figure 3.9 shows McArthur's Universal Corrective Map of the World, an upside down (or "down under") map drawn by some rebellious souls who apparently think Australia is the center of everything! Try thinking about how your assumptions about up-down-left-right would be different had you been taught this map in elementary school.

With the conventional biases of Mercator, travelers planning a trip from the United States to Europe often have a mental model of flying in an easterly, even rightward, direction. Some people are surprised to learn

Figure 3.8 *The Mercator map of the world is so familiar that it affects many of our everyday notions about the meaning of left, right, up, and down.* ·

Figure 3.9 *McArthur's Universal Corrective Map of the World, drawn by self-centered Aussies, shows me as a North American how self-centered my own Mercator-inspired notions are.*

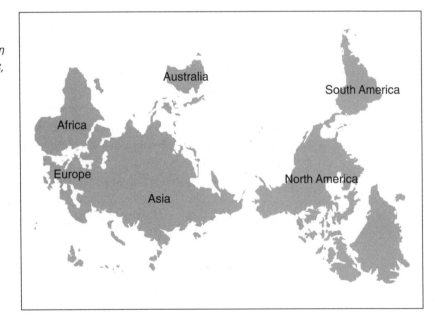

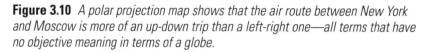

Figure 3.10 *A polar projection map shows that the air route between New York and Moscow is more of an up-down trip than a left-right one—all terms that have no objective meaning in terms of a globe.*

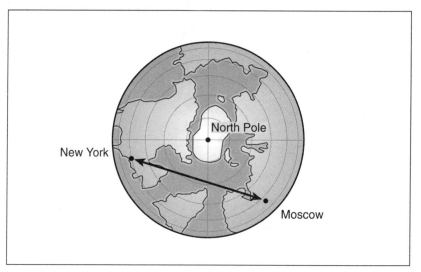

that the flight plan from most U.S. cities would follow a Great Circle route that goes "up," over the North Pole, as shown in Figure 3.10.

This cultural convention can also be seen in television advertising for air travel (see Figure 3.11) . In California, an airline that advertises flights between Los Angeles and Chicago shows an aircraft flying from left to right, with its nose pointed toward the upper right of the screen. Viewers in Los Angeles, who are thinking primarily about their departure rather than their return, perceive this screen direction as meaning *northeast*. It's fascinating to view similar spots that are prepared for Chicago audiences: As you might expect, the plane is pointed in exactly the opposite direction—southwest. In such cases, the art director is designing the scene to meet viewers' expectations.

The orientation of objects on a screen can even have visceral effects on an audience. Map conventions are so internalized that they influence involuntary eye and body movements. When you read a map, you have been taught to turn it so that north is at the top. Some of us will also go to the extreme of turning our bodies to face north while reading the map.

Figure 3.11 *In television, movies, and travel advertising, a journey to the right usually means* eastward, *left means* westward.

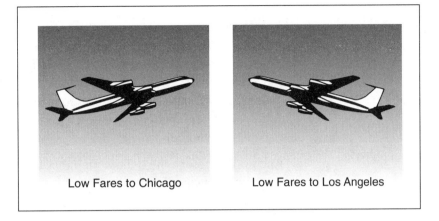

Low Fares to Chicago   Low Fares to Los Angeles

What's Wrong with This Picture?

Photographs or drawings that involve geography can also play to these prejudices about screen direction. Again, because of our dependence on Mercator, people in North America think of the Atlantic Ocean as being on the right and the Pacific on the left. If this seems silly to you, look at the sketches in Figure 3.12.

Figure 3.12 *Many people will think that the picture on the left is wrong, although they might not be able to tell you why. Because of Mercator, they have the notion that the Pacific Ocean should be on the left, as it is shown in the picture on the right.*

Sunny California   Sunny California

Lessons for Chart Makers

Again, distinctions of geography and direction are purely arbitrary. The real information content of airline commercials or the beach scene in the preceding examples is not affected by screen orientation. However, to the extent that people have deep-seated expectations, you should use them to your advantage.

And don't ignore them! If these prejudices aren't working for you, they are working against you.

For example, the chart in Figure 3.13 shows sales by region. The audience will be uncomfortable, and therefore needlessly distracted, because the West is not on the far left, Central in the middle, and East on the right of the *x* axis.

Don't get me wrong. It's not important to list the sales regions in any particular order for the sake of the honesty or integrity of the graph. The composition is just less effective because it ignores the prejudices of the audience. The chart might make them uncomfortable, even if the effect is only subconscious. And, because they are not focusing

Figure 3.13 *To meet viewers' expectations about direction and geography, the maker of this chart should have put the West on the left, Central in the middle, and East on the right.*

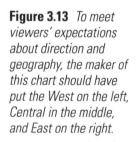

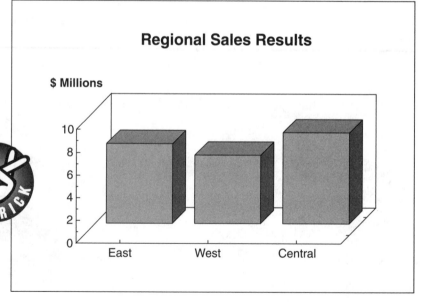

completely on the data, the chart's business message will suffer. The chart might not be a lie, but despite the presenter's good intentions, it's certainly an awful shame!

These notions about geography also affect other elements of your presentation, such as photographs of plant locations, scenic backgrounds used for charts, diagrams of computer networks, and presentation of demographics (statistics by population group and by geographic region).

One Picture Says It All

The graphic in Figure 3.14 summarizes this chapter by committing three serious errors. First, the increase in sales is shown as horizontal bars, sacrificing the intuitive power of up-down. Second, the vehicle in the picture is traveling from right to left, a retrograde path that conveys the *opposite* of a feeling of progress (not a good technique when you're talking sales).

Figure 3.14 *Guess what's wrong with this picture and win a trip to the next chapter!*

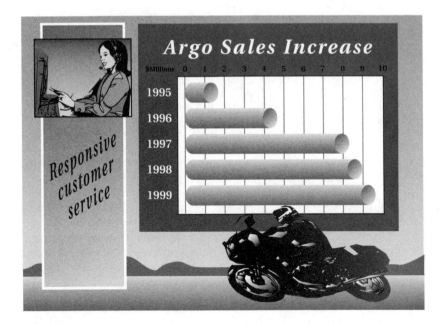

The third error is one I haven't covered yet, and it's actually a compound error. Notice that a photo of a customer service representative has been inset near the graph. Now, to us intelligent primates, human faces are the most interesting thing on earth (despite what you might think about attractive bodies with little or no clothing). The face, in itself, is a major distraction from the intended message, simply because it is the wrong place for the audience to be looking. Furthermore, the face in the photo is looking to the *left* (not good), and (worst of all) her gaze is off the edge of the chart, away from the real subject of interest—the sales data!

Consciously or not, when the audience sees this chart, they will look at the face first, and then look where the face is looking. In the moment it takes to focus back on the bar graph, the presenter might have made his point and moved on to the next chart! (What was that he said?)

The next chapter gets truly serious about *xy* charts, which, as I've said, capitalize on our prejudices about orientation.

XY Charts

Or, The Old Story of Money and Time

$\mathbf{P}$ERHAPS **the most common scheme** for business charts is a plot of quantity versus time, as shown in Figure 4.1. Amounts—of dollars, say—increase or decrease along a vertical scale, or *y axis*. Time progresses from left to right along a horizontal scale, or *x axis*.

Many people assume they know how to read these charts because this up-down, left-right arrangement is so intuitive. In Chapter 3, I describe how this *xy* plotting scheme relates to meanings we have given to

Figure 4.1 *People in Western countries seem to understand intuitively that up-down means* amount *and left-right means* time.

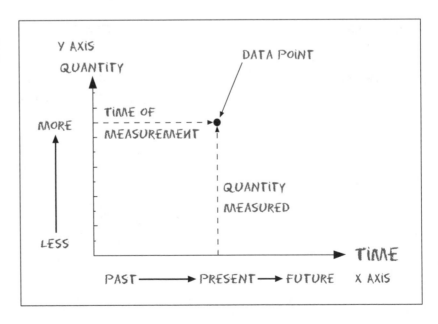

Thank You, René Descartes

The *xy* charting scheme is derived from the Cartesian coordinate system developed by the seventeenth-century French mathematician and philosopher René Descartes (whose name, by the way, can be translated "René of the Charts"). If you took trigonometry in school, you've probably had it with old René, but here's one of those dry subjects that can finally help you persuade people and make money!

In the Cartesian system, the position of a point in two-dimensional (2D) space is determined by two numbers, or *coordinates*, as shown in Figure 4.2. Each coordinate is a distance from a central point, or *origin*. A point in 2D, such as on a sheet of paper, can be located *x* numeric units horizontally and *y* units vertically from the origin. So drawing a line requires at least four numbers—two for each of its endpoints.

A 2D drawing or chart in the Cartesian system is flat, or *planar.* To add depth, a third dimension, or *z* axis, can be used, as shown in Figure 4.3. In 3D, a point in space is defined by three coordinates—*x, y,* and *z.*

In a different scheme called *2½D* representation, the *z* axis is used—not to add depth or solidity to a chart or drawing—but to hold separate drawing planes, like a stack of clear sheets (Figure 4.4), or *overlays.* In 2½D, the *z*-axis value is the *number of the sheet* on which an object is drawn. The order of the sheets determines the *visual priority* of objects. Objects on sheets with high visual priority cover up and hide the parts of other objects that lie directly beneath them.

Figure 4.2 *The Cartesian system used for most charting is two-dimensional (2D).*

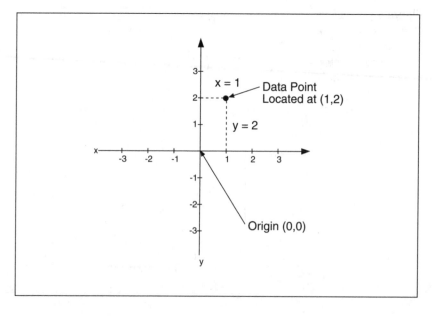

Figure 4.3 *Three-dimensional (3D) plotting of solids requires three coordinates for each point—x, y, and z.*

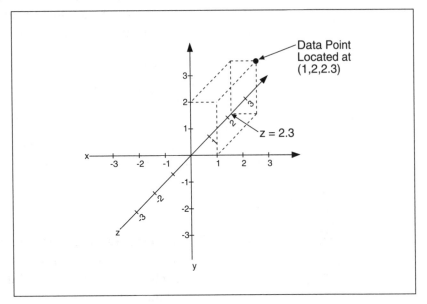

Figure 4.4 *Multiple-layered drawings can be represented in 2½D.*

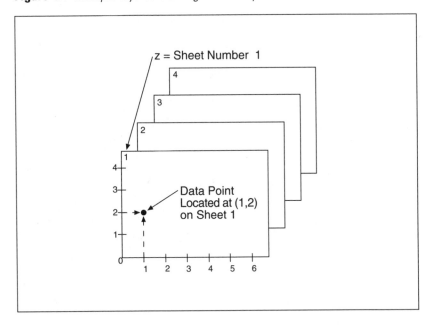

directions in space: Especially for Western audiences, upward motion means gain or increase, and motion from left to right conveys the progress of time. If Hebraic scholars—who read from right to left—had invented *xy* charting, a typical plot might look like this instead:

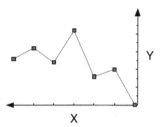

In the most common type of *xy* data chart used in business, a sequence of data values is plotted as a series of points on the chart in left-to-right order. The height of each point is its *y* value. Its *x* value is simply the next division of the *x*-axis scale:

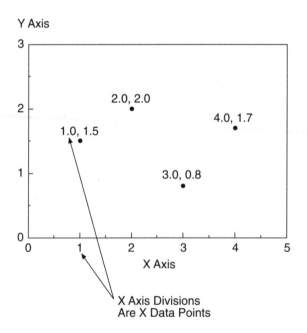

For most types of *xy* charts used in business, the numeric value of a data point is given as just the *y* value. The *x* values are the corresponding *divisions*—spaced at regular intervals—along the *x* axis. Often, x-axis divisions are marked by labels instead of numbers. For example, because the x axis usually represents the flow of time, the x-axis divisions might be months, labeled JAN, FEB, MAR, and so on. If you had three dollar amounts, such as 150, 185, and 210, these would be plotted in sequence as *y* values: 150 at JAN, 185 at FEB, and 210 at MAR. A chart type that does not necessarily locate points at regular x-axis divisions is the *scatter* chart, in which a pair of numbers (*x* and *y*) is required to locate each point.

The sequence of points is called a *data series*, or *data set*. Lines or shapes are drawn to link the points in whatever plotting style you prefer. In these different plotting styles, data points are shown as heights of separate bars (*bar chart*),

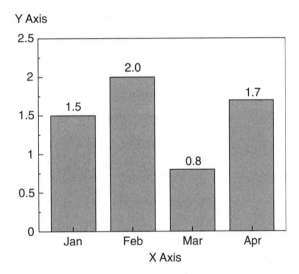

are connected to form a line (*line chart*),

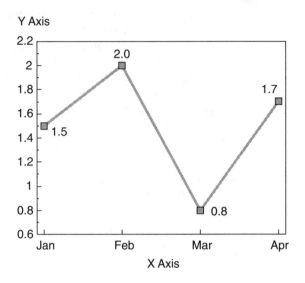

or are connected with shaded areas underneath (*area chart*):

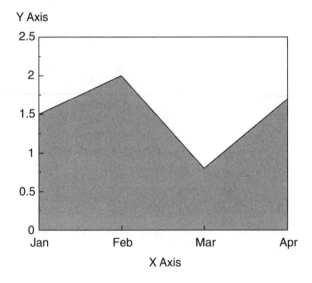

Data points can also be shown as dots or symbols in what's called a *scatter chart*, or *point chart*. In a scatter chart, the points aren't plotted sequentially. Two coordinates are used for each point, meaning that the *x* value can fall anywhere along the scale—not just at major divisions:

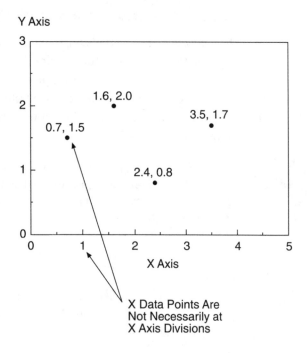

Scientists and statisticians—who labor meticulously to record each precious data point in lengthy experiments and surveys—just love those scatter charts. The idea is to throw the "raw" data up there and then look for patterns. The rest of us generally prefer something more digestible.

Some people use the term *xy chart* to refer only to scatter charts. In this book, I use it more generally to describe any of several chart types that have *x* and *y* axes, such as bar, line, and area charts.

These *xy* plotting styles can have many variations, including stacked and cumulative plots, and special formats.

Stacked or Cumulative?

In an *xy* chart, sets of data can be built on one another in two different ways: stacked or cumulative.

Stacked Charts Are a Bumpy Ride!

In a stacked area or bar chart, each data set uses the previous set as its baseline: Each *y* value is added to the maximum *y* value of the previous set. So, where one set peaks, the next begins. In effect, each data set rides the bumps of those below it—built upon one another like bricks and mortar or layers in a cake. A set of stacked areas is shown in Figure 4.5.

It's not nice to stack the *lines* in a line chart. The notion of stacking won't be visually apparent, as it is with solid areas or bars. By convention, the baseline of each series in a line chart is usually assumed to be the line *y*=0 (the x axis).

Figure 4.5 *Area charts are usually shown with stacked data series. That is, the starting point of one area is the top of the area below it. Putting the least "bumpy" data series at the bottom minimizes distortion of the other series.*

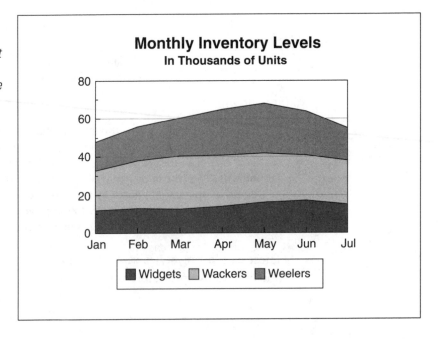

For bar charts only, an alternative to stacking several data series on the same chart is to *cluster* them. Separate bars are shown for each series at each division of the x axis (Figure 4.6).

Figure 4.6 *The clustered bars in the lower chart are an alternative to the stacked bars of the upper chart. Note the difference in the ranges of the y-axis scales—from 0 to 30 for clustered plotting, from 0 to 80 when the series are stacked on one another. Notice also in the stacked chart that the series with the least fluctuation has been placed on the bottom to minimize distortion of the other series.*

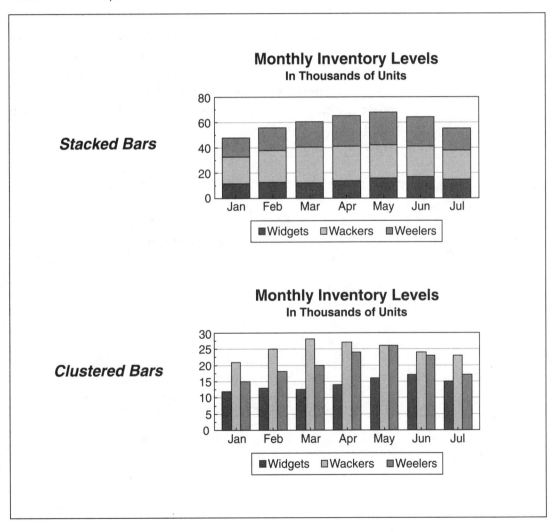

A clever liar can use this characteristic of stacked charts to his or her advantage: The fluctuations, or bumps, in the lower data series will enhance and exaggerate the bumps in the upper ones.

The bumpy effect will be most pronounced in area charts but can also affect perceptions about stacked bar charts. So, if you want to exaggerate the fluctuations in the data, put the bumpy series at the bottom (look back at Figure 4.6). Careful presenters who strive for honesty, however, will put the flatter data series on the bottom to minimize distortion in the other series.

As a general rule, area charts are plotted as stacked series, and line charts are not. Compare Figures 4.5 and 4.7, which both plot the same sets of data. In Figure 4.5, the areas are stacked, building on one another. The baseline of each series is the series beneath it. In Figure 4.7, which is a line chart, the series are not stacked, and each uses the *x* axis as its baseline.

Figure 4.7 *The baseline of all series in a line chart is assumed to be the x axis (y=0). (This chart plots the same data as Figure 4.5.)*

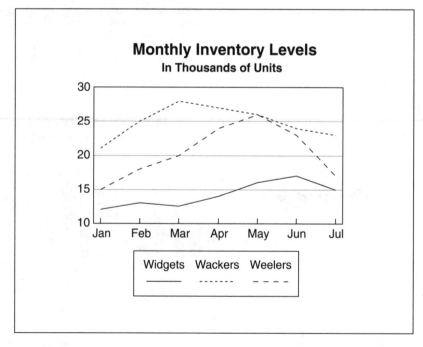

Cumulative Charts Can Be Confusing

Figure 4.8 shows a different method of plotting. Each y value in a set—in this case, the rainfall in inches—is a running total of all the preceding values in the *same* set. This is the correct sense of the term *cumulative* when you're talking about charts.

Cumulative charts are a favorite with liars because the general impression is always optimistic: As long as the data points are positive numbers, the trend is always upward—even if some of the increases are small. In the rainfall example, September, October, and November were fairly dry months in both cities, but that fact is obscured by the upward slope of the cumulative totals.

Don't confuse *stacked* and *cumulative!* Stacking data series *does* make them appear to accumulate on top of one another, but this is not the meaning of *cumulative*. In a cumulative chart, the data accumulate only *within each series*. And, to avoid really confusing yourself and your audience, never mix stacked and cumulative plotting in the same chart! (Read more about it in Chapter 7.)

Figure 4.8 *In a cumulative chart, each point within a data series is a cumulative total of all preceding points in the same series. Cumulative charts look optimistic by nature.*

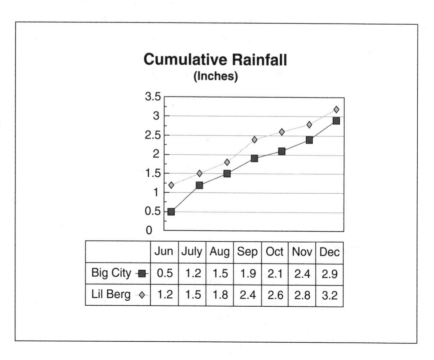

Cumulative Rainfall
(Inches)

		Jun	July	Aug	Sep	Oct	Nov	Dec
Big City	■	0.5	1.2	1.5	1.9	2.1	2.4	2.9
Lil Berg	◇	1.2	1.5	1.8	2.4	2.6	2.8	3.2

A careful presenter will only use cumulative charting for quantities that actually accumulate. For example, you could use a cumulative chart to show the growth of the principal in an investment with compounded interest. In such an investment, interest from the prior period is added to the principal amount before interest for the current period is calculated.

Special Charts for Special Occasions

Some special ways of plotting *xy* data are unique to specific fields or industries. These include *high-low-close-open (HLCO)* charts and *Gantt* charts.

Follow Stock Prices This Way

Stock prices are typically shown as *floating bars* in a format called *high-low-close-open (HLCO)*. Each bar shows the high and low prices for a stock on a given day. Open and close—the prices at the opening and closing of the day's trading session—are shown as small horizontal *ticks* in the bars. Here are two different styles of HLCO floating bars. The thin one on the left is also called a *bar*, even though it's just a segment of a line.

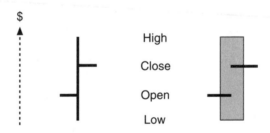

Figure 4.9 shows another HLCO chart. In this example, the number of shares of a stock traded each day is plotted as a separate set of bars at the bottom of the chart.

Figure 4.9 *In this HLCO chart, daily volume of this stock is plotted as another data series at the bottom.*

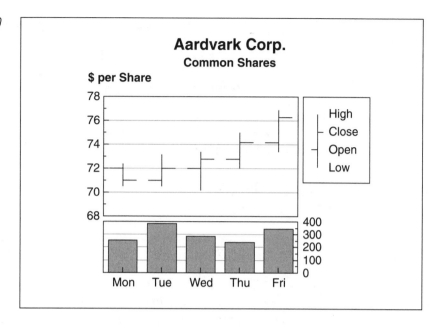

HLCO charts can be distorted by playing with the appearance of the bars (skinny or fat) and by manipulating the *x* and *y* axes. Warfare using axes is a big topic I'll defer to Chapter 6.

Project Managers Love to Stretch Mr. Gantt

As I point out in Chapter 3, another common use of floating bars is to represent time spans, or *durations*. In a *Gantt chart*, time flows in the familiar direction—from left to right—but the divisions of the *y* axis represent not quantities, but the different phases of a project or larger time span (Figure 4.10).

Project managers routinely use Gantt charts to show work schedules. Progress against a schedule can be highlighted by using two sets of floating bars: Planned and Actual. Any gap between the planned and actual bars represents a visible *slippage* of the work schedule.

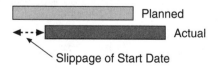

Figure 4.10 *In the Gantt charts used for project management, durations of work activities are plotted as horizontal bars.*

Software Development Project

ID	Task	Start	March				April					
			03/07	03/14	03/21	03/28	04/04	04/11	04/18	04/25	05/02	05/09
1	Design	03/15/99										
2	Review	04/05/99										
3	Test	04/19/99										
4	Debug	05/03/99										

Notice in Figure 4.10 how the interpretation of the floating bars is helped by the notion of time flowing from left to right. This impression would be defeated if the same durations were plotted instead as conventional, vertical bars, as shown in Figure 4.11. When the audience sees the vertical bars, the impression will be that the project is static, not making progress, even though the data in both cases is the same.

Project managers can lie with Gantt charts. If the *x* axis is elongated, the audience will get the impression that the overall time span has been lengthened. (All project managers want more time because they are always up against tight deadlines!) There's more about playing with axes in Chapter 6.

The Lesson of the Grasshopper and the Gantt

Gantt charts are named for Henry Gantt, an American industrial engineer of the early twentieth century. Gantt described business organizations as mechanisms that are subject to three basic managerial functions: planning, organization, and control. Features of his charting technique have been designed to support each of those functions. Gantt's theories were studied by the young W. Edwards Deming, whose own work in quality control was taken much more seriously in postwar Japan than in his own country, the United States. Some historians believe that adopting Deming's methods was a major factor in Japan's stunning success in high-tech manufacturing.

The lesson? Study the Gantt, thou sluggard, and grow wise!

Figure 4.11 *Plotting durations as vertical bars makes a project seem static.*

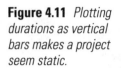

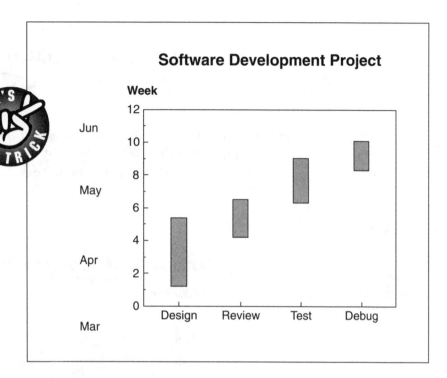

Hanky Panky with *X* and *Y*

There are some cardinal sins you may want to commit, or not, when designing *xy* charts. Here are two of the liars' favorite tricks: First, turn vertical charts into horizontal format, subverting the intuitions of the audience about quantity (up-down) and time (left-right). Second, mix proportional and quantity-time relationships in the same chart. For example, use symbols instead of bars, which you'll see is a really nasty trick! (For an example, look ahead at Figure 4.13.)

Laying and Lying

Worried about the highs and lows in a vertical bar chart?

Laying a vertical chart on its side can be a form of subtle lying because you confuse the intuitive meanings of up-down and left-right. Figure 4.12 shows the same chart in both the conventional vertical bar

Figure 4.12 *Turning a conventional vertical bar chart on its side confuses the intuitive meanings of up-down and left-right.*

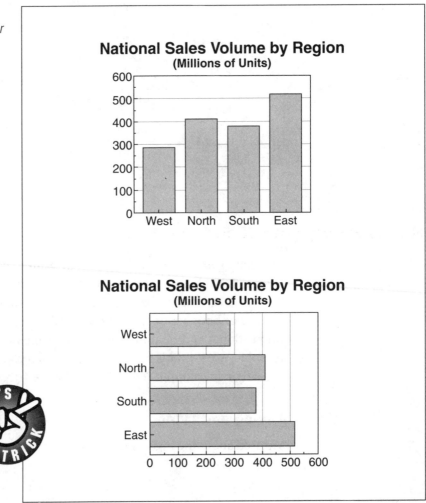

format and the *horizontal format*. Notice that the effect is quite different from Gantt charts, which also use horizontal bars but *preserve the flow of time from left to right*. In a horizontal chart, the positions of the axes have been transposed: The *x* axis—which usually represents time—runs up and down, and the *y* axis—which usually represents quantity—runs from left to right.

The horizontal format is misleading because the audience will need a moment—however brief—to adjust to the counter-intuitive layout. An accomplished liar can use this slight delay in comprehension to achieve nefarious purposes. For example, the unscrupulous presenter might say something in that moment about the chart that he or she doesn't want the audience to fully understand: "These numbers are hypothetical, of course."

Horizontal bars used to represent quantities are misleading for another reason. They will be perceived as if they were *dependent on time*, and progressing toward a goal. The longer bars, which would be taller on a conventional vertical chart, will be seen as *closer to the goal*. In reality, there might not be a goal, or the goal might not be relevant to the chart.

Honey, I Blew Up the Data!

Recall in Chapter 2 that I describe pie charts as a means of showing proportional relationships, such as percentages or ratios. *XY* charts, on the other hand, are designed to represent actual quantities.

It's a recipe for disaster to mix proportional and actual relationships in the same chart! The designer of the chart in Figure 4.13 apparently thought it would add a creative touch to use symbols of people instead of bars.

The chart with the people symbols is a *huge* distortion of the data! This distortion arises from the difficulty of estimating the areas of objects visually, just as with proportionally-sized pies. (The gender stereotypes in this illustration might also preoccupy the audience, but that's a different issue.)

In the example, the projected increase from this year to next is just 20 percent. The designer of the chart has made the people on the right 20 percent taller, as though they were bars. However, the shapes on the

Figure 4.13 *The people on the right are 20 percent taller (the correct comparison) but are* 200 *percent larger* by area!

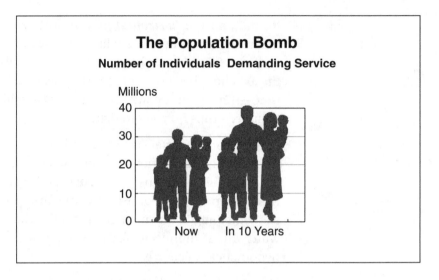

right—which have increased in width as well as height—are actually *twice* as large, by area. In other words, the increase is made to look much bigger than it really is: *200 percent instead of 20 percent!*

Tricks of Perspective

A favorite trick of liars who hang out in bars is to use 3D to make all the bars in the chart appear taller. In general, there's a right way and a wrong way to add depth to a bar chart. The difference lies in a trick of *perspective*.

In a 2D perspective drawing, the sides of objects appear tapered, converging toward one or more points on the horizon called *vanishing points*. A common type of perspective drawing uses two vanishing points, one at each end of the horizon:

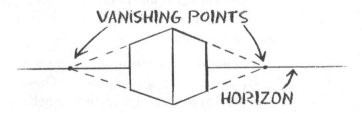

Figure 4.14 *Careful presenters put vanishing points* below *the tops of 3D bars.*

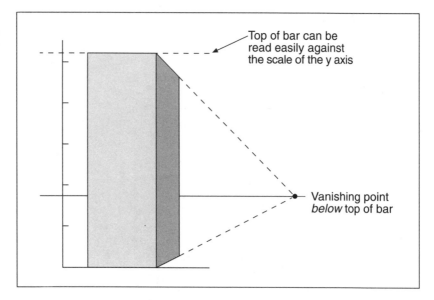

The rule for drawing perspective bars is simple: Careful presenters put the vanishing points *below* the tops of the bars, as shown in Figure 4.14. This makes the "front face" of each bar its tallest part, and it's obvious where the top of the bar is in relation to the scale.

The liar's trick with 3D is to put the vanishing points *above* the tops of the bars. This emphasizes the back edge of each bar, which appears higher than the front. This technique is a trick of perspective because the scale division also runs diagonally: The lower front part of the bar is just as tall as the back part. The trick makes it more difficult for viewers to estimate the heights of the bars accurately. The effect is shown in Figure 4.15.

If you plot multiple sets of 3D bars on the same chart, you might have no choice but to place the vanishing point above the tops of the bars, creating the potential for confusion shown in Figure 4.16. If you put the vanishing point lower, the bars in back might not be visible. Some computer graphics programs draw the bars this way automatically, presumably because it reduces the amount of overlap when you have multiple sets of bars. If you place the vanishing point above the tops

Figure 4.15 *It's a liar's trick to put vanishing points* above *the bars so that the back edges appear higher than the front.*

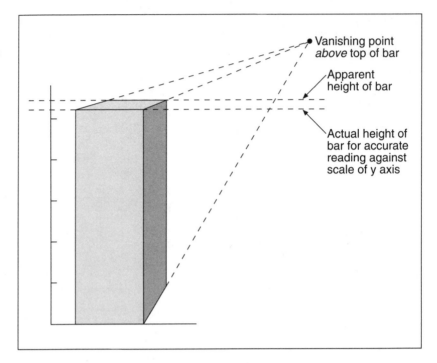

Figure 4.16 *In this set of 3D bars, the vanishing point is* above *the bars, making it difficult for the audience to estimate the true heights.*

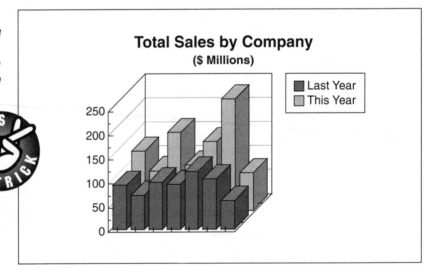

of the bars, be aware that the audience will have difficulty estimating the true heights of the bars. Consider showing the numeric data as labels at the tops of the bars or in an accompanying table.

An Exception to the Golden Rule of *X* and *Y*

In an *xy* chart that strives for honesty, the vertical *y* axis should be an amount and the horizontal *x* axis should indicate the flow of time from left to right. To do it otherwise invites misunderstanding, *with one exception.*

In some bar charts, such as the one in Figure 4.17, divisions of the *x* axis are not time spans, but labels of different categories.

LIAR'S TRICK Even if the *x* axis has nothing to do with time, the audience won't be able to avoid associating left-to-right movement with *progress*. If you want to exploit this tendency, put the set of bars you want to emphasize on the *right end* of the chart. Liars may want to shift older, perhaps more favorable, data to the right end of the *x* axis.

Figure 4.17 *Even if the x axis does not represent the flow of time, viewers will regard the bars on the right as the most recent or important. That's not necessarily bad, just something you should be aware of.*

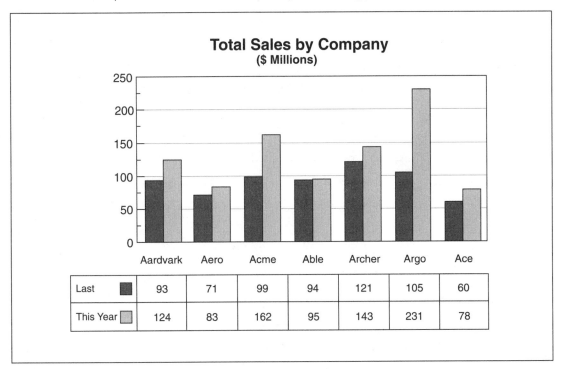

Total Sales by Company
($ Millions)

		Aardvark	Aero	Acme	Able	Archer	Argo	Ace
Last	■	93	71	99	94	121	105	60
This Year	▨	124	83	162	95	143	231	78

Looking Ahead to Your Future as a Zen Master

The next chapter covers an Eastern cousin of the *xy* chart:the *radar chart,* which looks like a mandala, if you want to get spiritual about it. You may be surprised to learn that for the same reasons that Western audiences prefer the familiar *xy* chart, the multifaceted radar chart is a great persuader in Japan!

Radar Charts

Or, Watch Out for That Zen Speed Trap!

A GRAPH that looks like a geisha's parasol without the silk is called by the less picturesque name *radar chart*—a term borrowed from the target-shaped screen used by air-traffic controllers.

I can't resist the image of the Japanese parasol, because I've heard that these radar charts are particularly popular in Japan. The reasons have much to do with the differences between the expectations of Eastern and Western audiences—but let's discuss the basics first.

Axes, Grids, and Criteria

Figure 5.1 shows a typical radar chart—a rating of a new car model according to five criteria: fuel economy, handling, acceleration, styling, and ride. Like the spines of an umbrella or the spokes of a wheel, several axes radiate outward from a common center. The optional grid lines are shown as concentric circles. Each grid line marks a uniform distance from the center. In this case, the distance from the center on a grid line is the car's score for that criterion.

Each axis can represent a separate type or category of measurement. (The mathematicians call it *multivariate analysis*.)

If you want to be able to draw conclusions from the *shapes* of the data plots, the scoring must be consistent for all criteria. For example, let's say you rated a company's customer service on promptness, courtesy, accuracy, effectiveness, and follow-up. So, if you score courtesy on a scale from 1 to 10 using precision to tenths of a point (6.8), you should score the other criteria exactly the same way. Liars don't necessarily follow this rule, for reasons that should be clear at the conclusion of this chapter.

Figure 5.1   *The axes and grid lines of a radar chart let you describe a thing graphically according to multiple criteria—all in the same chart. In this case, a new car model is evaluated on a scale of 0–10 in five criteria: fuel economy, handling, acceleration, styling, and ride.*

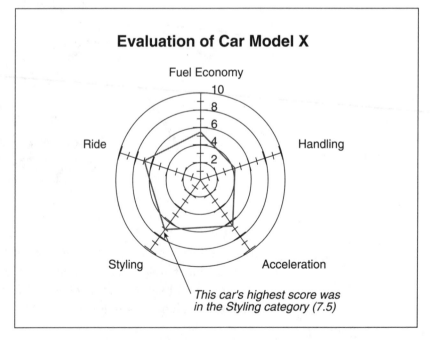

A Delicious Example

This multivariate stuff is best explained by example. Continuing in the Japanese mode, let's assume that I'm the emperor's food taster. It's my job to sample the emperor's meal before he takes a bite—to make sure he isn't poisoned, yes, but also to advise him whether a dish is appetizing and worthy of his discriminating palate.

In my long and lucky years as the emperor's most trusted surviving taster, I have developed the habit of rating each new dish based on five key criteria: appearance, aroma, taste, texture, and digestibility.

In each of these categories, I give each dish a numeric rating from 0 (disgusting) to 10 (most worthy). Here is a recent memorable meal, which I have set out as an attractive table.

	APPEARANCE	AROMA	TASTE	TEXTURE	DIGESTIBILITY
MISO SOUP	6.2	5.9	6.0	6.1	6.2
CUCUMBER SALAD	4.5	2.0	4.2	6.5	6.0
TUNA SURPRISE	7.0	6.5	5.3	3.2	4.1
COOKIES	3.4	9.5	9.8	5.6	2.6

Each of the dishes, in this case, is a data set. Figure 5.2 shows how one of the data sets—Miso Soup—can be plotted on the same radar chart grid you've already seen. The numeric values in the data sets are plotted as data points *on* each axis (rather than between the axes, as in an *xy* chart). For example, the Soup course is marked at 6.0 on the axis labeled Taste. Points in the same data set are then connected to form a line plot. So, line segments connect the Soup scores among the five axes. For Soup, the result is a nearly regular pentagon because the rating for each category is nearly the same.

The outermost plots have the highest scores. You can see in the table that Cookies got the highest scores in Aroma and Taste. Notice,

Figure 5.2 *The plot of the data set Miso Soup is a nearly regular pentagon because it scored equally well in all criteria.*

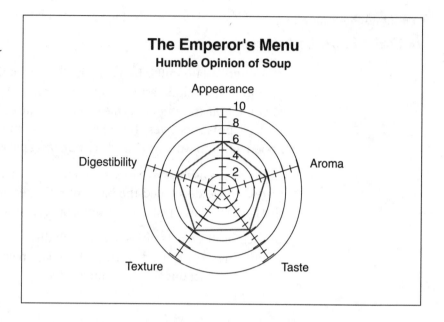

Figure 5.3 *The plot of the data set Cookies is irregular because it scored high in only two criteria— Aroma and Taste.*

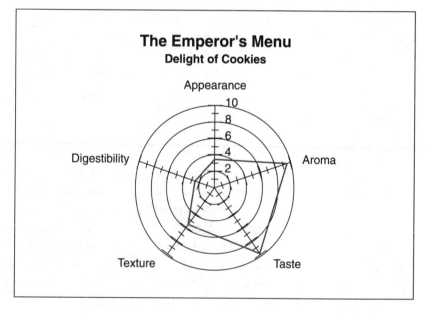

however, that the Cookies plot in Figure 5.3 is irregular in shape. Its pentagon looks squashed: Low scores in Appearance and Digestibility distort the plot.

Figure 5.4 *The shapes—and merits—of the dishes can be compared when you plot them all on the same chart.*

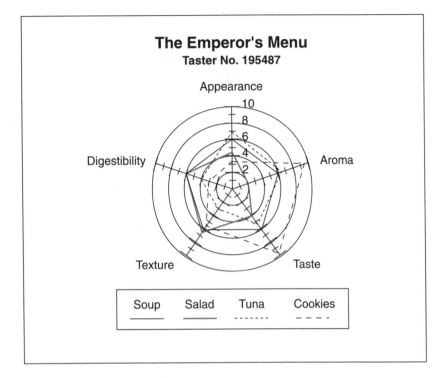

The soup scored much lower than the cookies in Taste and Aroma, but it had uniformly good scores in the other categories. As a result, its plot is regular, or symmetrical. Soup is therefore not the most spectacular, but it is certainly the most "well-rounded" dish, both in terms of overall scores and its symmetrical plot on the radar chart.

The scores of the various dishes can be compared at a glance by plotting them all on the same set of radar axes, as shown in Figure 5.4.

A Zen Quest for Symmetry

The quest for symmetry is the beauty of the radar chart. Its format implies a desire for balance and harmonious composition. The data set with the most nearly regular shape will be the most well-balanced among the various criteria. Regular shapes therefore represent optimized choices, equalized trade-offs, best overall values. By contrast, the more distorted the shape of a radar plot, the more it will be skewed in favor of one or more of the criteria, and so will have a less desirable mix of scores.

So, the radar-chart format imposes a bias on the interpretation of the data: Symmetrical plots are "good," asymmetrical "bad." This bias is just as strong—although decidedly different from—the up-down, left-right biases inherent in xy charts.

Herein perhaps lies the reason for this chart's popularity with Japanese business managers. The conventional xy chart is rooted in the Western notions of progress (x axis goes left to right) and gain (y axis goes up and down). The radar chart, on the other hand, is literally centered. The impression is static, unhurried. The objective is balance, consensus, harmony. In this scheme of plotting things, the spectacular but erratic performer will always look a bit odd. The most pleasing shape will be presented by the competitor who is perhaps only adequate, but is nevertheless consistent in all areas.

Because the managers seek harmony and balance, the radar chart suits their mindset. They will be prone to reject alternatives that don't look balanced.

The Hazards of Navigation by Radar

In an honest radar chart, all axes should have the same scale. If axes have different scales, the symmetry of the plots, if it appears at all, will be meaningless.

Cheaters might pretend that the dissimilar scales on their radar charts are proportional to each other. For example, three spokes of a mixed-up chart might represent different scoring methods: 0–4, 0–100, and 1–10. If the scales are truly equivalent, you should convert all scores so they can be plotted against comparable scales. If, however, there is really no relationship between the different scales, you shouldn't attempt to use them on the same chart.

Let's say you have two sets of criteria: One scale goes from 0 to 4, the other from 0 to 100. It might seem perfectly honest to stretch the first set of scores to match the second so that both sets can be plotted against the same scale. That is, you'd take a score of 4, which is at the top of the 0–4 scale, and *scale* it upward, or expand it proportionally-multiplying by 25. The result is 100, the same as the top of the 0–100 scale.

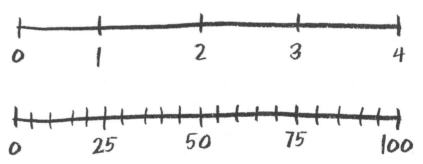

However, the scores on the 0–4 scale are probably less precise than those on the 0–100 scale. You can see this if you try the conversion in reverse: Convert the 0–100 set of scores to the 0–4 scale by dividing by 25. Let's say you're scoring to the nearest tenth, or one decimal place, and you have a score of 85.6 on a scale of 0–100. Divided by 25, that

85.6 score becomes 3.424 on the 0–4 scale. Now, it's unlikely that you would have taken the pains to refine your scores to *three* decimal places on the 0–4 scale, so the two sets of scores still aren't truly comparable.

As a rule, you should convert all sets of scores in a radar chart to the *least precise* scale. In this case, you would divide the 0–100 scores by 25 and plot them on the 0–4 scale rather than scaling the 0–4 scores upward. *You must then round all sets of data to the same number of decimal places.* So, if you convert 85.6 to 3.424, you must round the score to 3.4 so that it is comparable with the other 0–4 scores. You will lose the fine gradations of the 0–100 scores, but you never had them at all in the 0–4 set!

To be fairer yet, you should *truncate* the values, simply lopping off the extra digits. For example, a result of 3.467 rounded is 3.5, but truncated is 3.4. Using the truncated values makes a more valid comparison because the finer precision can't even be estimated in the coarser numbers, where it was never measured in the first place.

Here's Another Radar Trap!

A particularly nasty trick with radar charts is to intermix valid charts with questionable ones as adjacent pages or slides in the same presentation—that is, a liar might show some radar charts that have consistent scales alongside some that don't. The impression will be that all the charts are internally consistent.

The most dangerous liar's tricks with radar charts come down to scaling the data to fit some preconceived comparison—to make a plot seem either more regular or more distorted in relation to other data sets on the same chart.

But the tricks don't end there. Liars will try not only to scale the data, but they might also *scale the scales!* A clever chartmaker can adjust and otherwise monkey around with the axes of a chart, the very device us honest types use to interpret the values of data points!

Figure 5.5 *Here's the same data I showed you in Figure 5.1, but with the scales adjusted to improve the shape of the plot.*

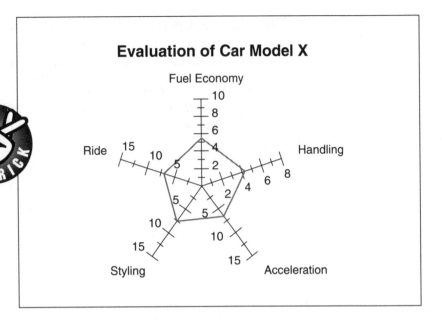

For example, I can manipulate the scales of a radar chart to make one of the plots seem more symmetrical, and hence, more desirable.

Look what I did in Figure 5.5. I took the data from Figure 5.1 and replotted it—this time, scaling each axis separately. Instead of requiring the scales to be consistent—each ranging from 0–10—I've set some of them to different maximum bounds. In my new chart, the maximum extent of any scale is always *two divisions greater* than its data value. (There's no special reason for picking *two*—the choice just happens to work well with this chart.)

Remember—neither the data nor the scoring method has changed at all: The scores still range between 0 and 10. However, some of the criteria are now scaled 0–15, another 0–8.

Maybe because I can sound sincere, my audience might think I have some good reason for doing this, such as *weighting* the scores. Weighting multiplies scores in incompatible units by different factors so that they can be compared on the same scale. But the awful truth is—I have no legitimate reason at all for adjusting the scales!

My dishonorable reason for changing the chart was just to make the plot prettier. It is now more regular in shape, and, according to the hidden biases of radar charts, this car now looks like a better buy!

To shield you against such heinous deception, I cover manipulation of axes—a subtle vice that can corrupt *xy* and radar charts alike—in the next chapter.

6

Axes

Or, Handy Tools for Dismembering Your Enemies

As wielded by chartmasters, the singular of *axes* (pronounced *ak*-seez) is *axis*—not *ax*. But in the wrong hands, these axes can be just as dangerous. On an *xy* chart, an axis is one of two numeric scales on which the chart maker plots data points. Axes also help readers of a chart estimate the numeric values of its plotted lines, bars, or areas.

Be on your guard! Manipulated by an ambitious liar, chart axes are no crude weapon. Their potential abuses comprise a whole armory of hazardous devices. Learn about them now, or you risk being surprised under battlefield conditions!

To understand these techniques, you must learn to think like the enemy. Whether your actual battle statistics are grim or hopeful, you can control what your audience thinks about a chart just by changing its axes. You can adjust levels of panic or enthusiasm about your message almost as easily as those Roman emperors who sent gladiators to their fates with a brusque thumbs-up or thumbs-down.

Now, you can't always turn a negative into a positive—that favorite trick of every salesperson who gets a customer objection. But you can either exaggerate or minimize a chart's net effect. You can smooth a wildly fluctuating cost curve until it seems barely a ripple. Or you can crank up a modest increase in earnings so that it skyrockets—literally off the scale.

I strongly advise you to become proficient with the chart warrior's gallery of weapons. You might not feel obliged to use them, but being well prepared might discourage your opponents from using them on you!

Some of these tricks apply to radar charts, as well. *XY* charts are described in Chapter 4, radar charts in Chapter 5. This chapter discusses *xy* charts, because the examples will be clearer in this commonly used format. But remember that these types of manipulation can be applied to any chart type that has numeric scales. For an example of radar-chart scales that have been seriously abused, see Figure 5.5 in the previous chapter.

Pick Up Those Axes and Follow Me!

Now that I've cautioned you about the harm you can do, let's take a closer look at the use and abuse of chart axes. Strictly speaking, an axis is a line that indicates a direction in space, or *dimension:*

Origin ●————————————→ Dimension X

Presumably as a help for chart readers, the line can be marked off at regular intervals, or *divisions:*

Marks that indicate divisions are called *ticks.* As a further aid to measurement, big ticks can mark *major divisions,* little ticks *minor divisions,* much as you'd find on a ruler:

Of course, the ticks don't mean much unless you label their values. An axis with labeled divisions is called a *scale:*

The minimum and maximum values on the scale define its *bounds.* All the values in between make up the axis *range.*

Those clever mathematicians have a special notation—a type of *subscripted* variable—for axis bounds. Referring to the *y-axis*, the minimum value is written y_{min} and the maximum y_{max}. The values in a data series, when shown as variables, can be numbered sequentially by the same method: y_1, y_2, y_3, and so on. If you don't know how many values are in a series, the last value is the y_n or *n*th.

Okay, let's see what kind of trouble you can get into by adjusting the scales on a chart.

Bounding the Range Is a Good Exercise

One way to draw an axis is to begin just a bit lower than the least value in the data series that you intend to plot, and to end a bit higher than the greatest value. For example, if the *y* values in the data series ranged from 3.5 to 8, it might seem reasonable to start the *y*-axis scale at 3.0, and end it at 9.0, as shown in Figure 6.1.

Notice that the trend in this plot peaks quickly, then moves decidedly downward. Because it measures performance—the average number of orders taken per month by each salesperson—the audience might

Figure 6.1 *The range of y-axis values is 3.5–8, and the bounds of this y-axis scale are 3.0 and 9.0.*

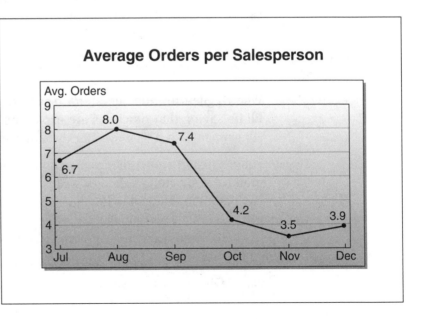

Figure 6.2 *Expanding the y-axis range here minimizes fluctuations in the data plot.*

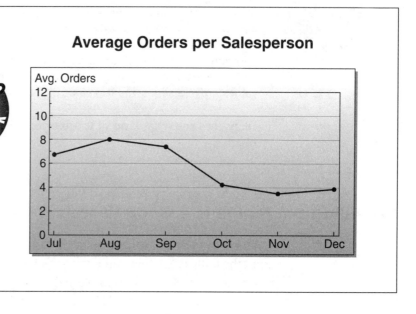

grow gloomy watching the line plummet. But, let's say I'm one of those unscrupulous presenters. I want to soften the harshness of my bad news—and I'm not reluctant to cheat a little.

On my revised chart, I'll start the *y*-axis scale at zero—that's logical, isn't it?—and end it at 12. Here's the effect, as shown in Figure 6.2. While I'm at it, I'll omit the data labels.

LIAR'S TRICK To flatten a plot—minimizing its fluctuations—*increase* the *y*-axis range.

Whew! I guess my numbers aren't so bad, after all. As I show them this, I'll be saying that our sales are experiencing a "modest seasonal adjustment"—and I'll move quickly to the next chart!

The next chart shows the average number of sales orders taken per month by my staff. Although we're not the top performers in the company, I'm thankful to note that there's been a modest increase—from 4.5 to 4.7. I had good luck starting the *y*-axis range at zero before, so I'll make the scale from zero to 5, as shown in Figure 6.3.

Unfortunately for my future with the company, this plot looks awfully flat. Maybe starting at zero wasn't such a good idea. I'll make the axis

Figure 6.3 *Here's another example where expanding the axis range flattened the plot. Unfortunately, that wasn't the effect I wanted.*

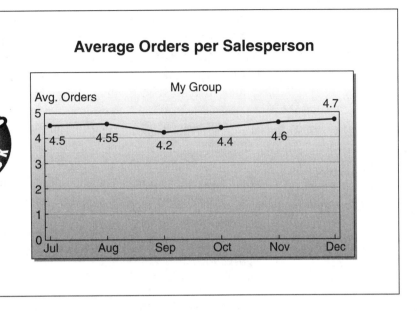

range narrower—*really* narrow this time—starting at 4.15. And I'll end it at 4.6. My last data point will go right off the scale, as shown in Figure 6.4. (Again, I think I'll leave those helpful data labels *off!*)

Figure 6.4 *Making the axis range narrower exaggerates fluctuations in the plot, which suits my deceitful purposes. I'm also pleased that the high point of the plot extends beyond the maximum value on the scale, which unnecessarily exaggerates the magnitude of that data point.*

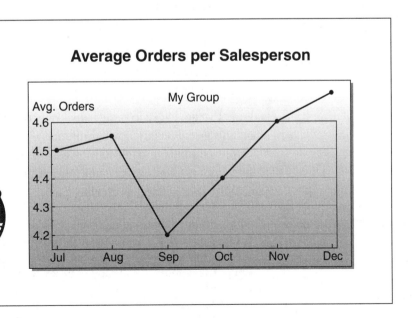

What an improvement (at least, from my deceitful point of view)! But, notice something else about my clever chart: Expanding the scale emphasized the valley, as well as the peak. True, it now looks as though my team is making a big comeback, but the audience might also wonder why we took that big plunge in September.

LIAR'S TRICK To emphasize a plot—maximizing its fluctuations—*decrease* the y-axis range.

What, you might well ask, is the "honest" way of plotting the data? The most truthful version of any chart will be the one that has the least deliberate distortion. Figure 6.1 isn't bad because its y-axis range is a typical one for the data being shown. But to be entirely fair to your audience—you should show that the y-axis scale doesn't start at zero, as described below in "Fighting Fair with Unconventional Axes."

Real Movers Don't Start at Zero

I may be a shameless liar, but I'm no fool—I don't want to get caught. I started the *y-axis* of the first chart at zero, the *y-axis* of the second at 4.4. The scales aren't consistent—they have different starting points and span different ranges. If I show the two charts one right after the other, some bright observer might spot my trick.

To hide my intentions, I'll try yet another trick. Here's what I'm thinking: *Unless I take pains to show them otherwise, the audience will usually assume that axis scales begin at zero.* Therefore, I won't label the non-zero starting point on the scale.

Wait! I've got an even better idea. Why don't I omit all but the top label on the *y-axis* scale? If the audience thinks that the scale starts from zero, I will have given the impression that the *rate of increase* in the sales plot is even bigger, spanning the range 0 to 4.7 instead of 4.2 to 4.7. The result is shown in Figure 6.5.

LIAR'S TRICK If it suits your purposes, don't label the starting point of the y-axis scale. The audience will have to guess where the scale begins, provided that they are attentive enough to even try. (Maybe they won't have time, if you talk fast enough and move quickly to the next chart!) If you want to be *really* nasty, don't label any of the scale divisions except the top one. The wider the range and the more divisions it has, the easier it will be to get away with your lie. For example, if the axis goes from 200 to 800,

Figure 6.5 *Omitting most of the y-axis labels achieves a minimalist graphic design that is nonetheless grossly misleading.*

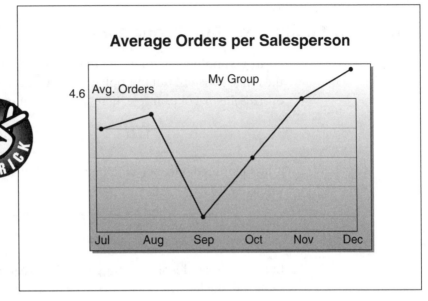

label only the top division at 800. The audience probably will mistakenly assume that the scale goes from 0 to 800.

Fighting Fair with Unconventional Axes

Even if you have some legitimate reason for not starting an axis scale at zero, there is a way to fight fair so that the audience can see you're not up to any tricks. Start the scale at zero, then show a break in the axis line before continuing the scale. And if you want to be unmistakably clear, run the break across the full width of the chart:

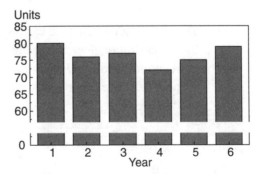

And don't make the scale labels too small!

Advanced Scaling for Ambitious Climbers

Scaling is another technique for manipulating axes. It involves keeping the range constant but multiplying or dividing its values by a number, or *scale factor.* For example, here's a scale that ranges from 0 to 4,000:

A less cluttered, cleaner-looking scale will result if you divide all its values by 1,000:

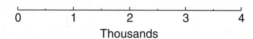

But notice that, despite the ranges of the two scales being exactly the same, the single-digit values on the second scale tend to de-emphasize the magnitude of the data.

Closely related to axis scaling is the whole issue of labeling (or mis-labeling, as liars tend to do), which is a major topic in Chapter 10.

This trick of manipulation is kept honest by the scale label—Thousands. However, the trick appears more deceitful when the label is positioned further away from the axis and perhaps also reduced in size, as in a less obvious footnote:

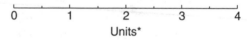

The magnitude of the data can also be either emphasized or minimized by increasing or decreasing the number of scale divisions and their corresponding labels. For example, of these two scales, the axis on top emphasizes the scale magnitude more—even though their ranges are exactly the same:

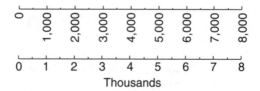

The trick of increasing or decreasing the number of scale divisions can also affect how the audience perceives the *precision* of your data. For example, a scientist who is presenting experimental data could give the (perhaps misleading) impression of precision by including both major divisions and very fine minor divisions:

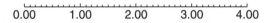

In reality, the data points might only be accurate within a fairly broad range—say, plus or minus one whole number (±1.0). However, dividing the scale into tenths would imply that the accuracy of the data points is much better—say, ±0.1.

For Extra Credit: Dueling with Dual Y-Axes

There are occasions—legitimate ones—for showing plots of two widely different magnitudes on the same chart. For example, if my company's labor costs are on the increase, I might want to show that curve as compared to U.S. Social Security receipts—a big factor in labor costs. My objective is to show that, although our costs are increasing, some huge trend that is outside my control might be pushing them up (so it's not really my fault!).

The basic problem, though, is that my company's sales are in millions of dollars, and the country's GNP is measured in trillions.

Plots of different magnitudes can be shown on the same chart by using two *y-axes*, with one scale on the left and a different scale on the right, as shown in Figure 6.6.

There's nothing inherently dishonest about such dual-*y* plotting. But as with the other techniques I've shown you, this one has its abuses.

Of course, there *is* a causal relationship between the expenses of maintaining the Social Security system and a company's labor costs. However, for purposes of my presentation, matters of national social policy are probably not at issue. My managers want to know what I

Figure 6.6 *A chart with dual y-axes can be used to compare plots of widely different magnitudes.*

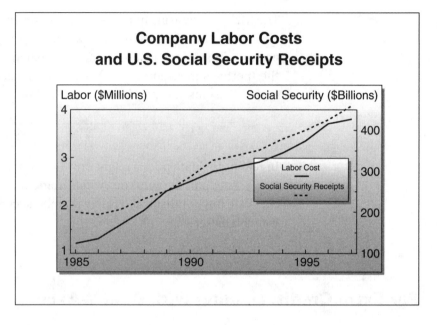

Figure 6.6 *A chart with dual y-axes can be used to compare plots of widely different magnitudes.*

can do over the short term to control costs. Even though the national economic trend probably has some effect on my situation, such large issues just aren't relevant to my presentation.

To be both clear and honest in a dual-*y* chart, show just two data series—one for each *y-axis*. As a further help to your audience, show a legend or color-code the plots to their respective axes. For example, use yellow for one plot and its left *y-axis* scale, green for the other plot and its scale.

A clever liar might also use dual-*y* plotting to make bogus comparisons. Look at the chart in Figure 6.7. Sales volume is plotted against the left vertical axis, the dollar-value of inventory on the other. The implied comparison is that the sales results are affected by the availability of product. This could be a valid comparison, although the truth is probably the other way around—that inventory is restocked depending on how fast the products are selling.

It's also possible that the lying chart maker knows perfectly well that the decline in sales has nothing to do with inventory. The liar might be trying to hide much more significant causes, such as cutbacks in sales staff or advertising.

Figure 6.7 *It's not necessarily incorrect to show dissimilar data sets on the same dual-y chart. But is the comparison valid?*

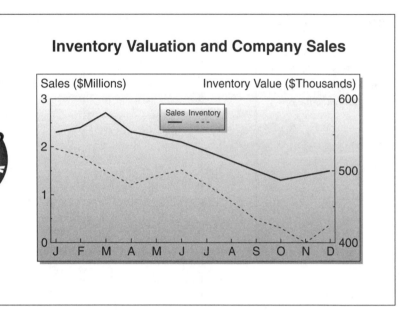

The Big Number Crunchers Fight with Logs

One of the more unusual tricks you can try with scales is to make them *logarithmic*. Logarithmic axes, or *log* axes, are scaled by powers of 10 (also called *orders of magnitude*). The legitimate purpose is to be able to fit a very wide-ranging data set on a single, compact chart. There really is no practical alternative to log scales—for example, if the points in the same data set range from the tens to the millions.

Although log scales in *xy* charts have their uses, this type of charting will not be easily understood by general audiences. Logarithmic plotting is used mainly by scientists and engineers, who typically deal with wide-ranging data, such as astronomical distances.

The most common (and most readily understood by nontechnical audiences) type of log plotting is called *semi-logarithmic*, or *semi-log*. The scale of only one axis—usually the *y*-axis—is logarithmic. The other—the *x*—is normal, or *linear*, as shown in Figure 6.8.

Both charts in the Figure plot the same data, the powers of the number 2: $2^0=1$, $2^1=2$, $2^2=4$, and so on. The chart on the left is the normal kind—

Figure 6.8 *A semi-log chart can shrink a wide-ranging data set down to size, but this plotting style might be unfamiliar to general audiences.*

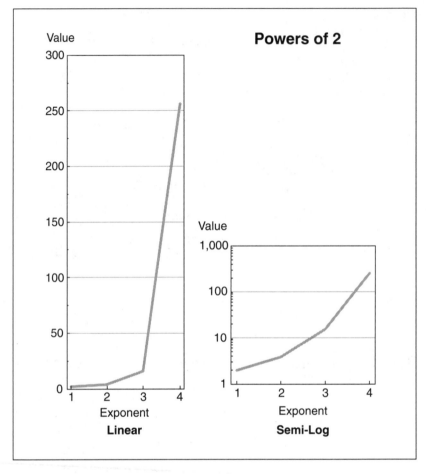

called *linear* because none of its axes is logarithmic. The one on the right is semi-log: The *y*-axis is logarithmic and the *x*-axis is linear. (More about a third type, log-log, shortly.)

Notice in Figure 6.8 that the spacing of the horizontal grid lines on both charts looks normal—until you check the *y*-axis labels. Here's another opportunity for mischief: The spacing of the grids can *appear* linear. Unless they study the scale labels, an audience won't guess right away that the chart is logarithmic!

Now, why would a liar want to use log scales?

In mathematics, wide-ranging trends that increase by powers are called *exponential.* Plotted against conventional linear scales, an exponential

Figure 6.9 *An exponential trend can be made to look less dramatic by plotting it on a semi-log chart, where it resembles a linear trend.*

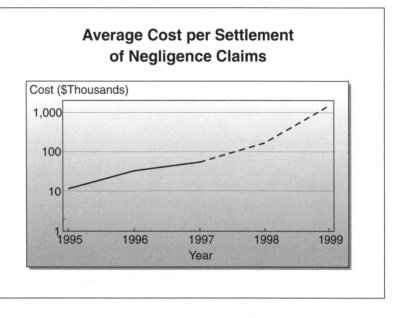

curve can scoot upward at a shocking rate, as shown on the chart on the left in Figure 6.8. However, plot the same curve on a semi-log scale, as shown in the chart on the right, and it begins to look more like a straight line, which can be much less worrisome. An audience unskilled in interpreting log charts might well be led astray!

There's a practical example in Figure 6.9. Here the presenter is attempting to make an alarming increase in costs look like a less dramatic linear trend.

It's a cliché of modern media coverage that "exponential" costs/crime/environmental impacts are "threatening to spiral out of control." Despite the facts that the actual trends might not be truly exponential or that exponential trends don't spiral, in popular usage the term *exponential* often conveys the notion of looming crisis. For people who might otherwise worry about these things, there's the false comfort of the semi-log scale!

As shown in Figure 6.10, a doubly obscure charting type called *log-log* uses log scales for *both* axes, *x* and *y*. Don't resort to this technique unless you are sure that your audience is used to seeing data plotted this way. But if you do get out the double logs, be sure to use grid lines on both scales—and label them clearly!

Figure 6.10   *Save log-log plotting for the propeller-heads. Nontechnical audiences may find it bewildering.*

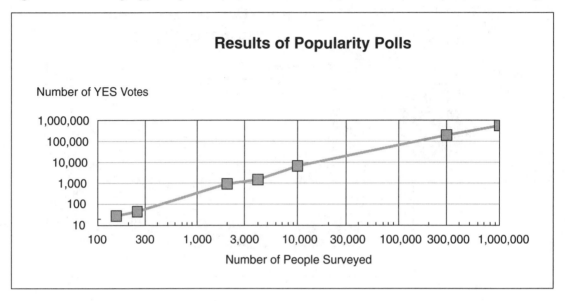

Your Mission, If You Choose to Accept It

It should be apparent from this brief survey of potentially lying tricks that you can make an *xy* chart say just about anything by adjusting its axis scales. If you had any preconceived notions that charting is an objective science, these should be dispelled by now! Ultimately—and this should come as no surprise—you have to decide whether a chart you have made correctly conveys your interpretation of the data. There's no pretension to *truth* here. Charting is a medium of communication, and I simply want you to communicate what you intend. And I'm not so much worried that your intention is to lie through your charts. The real tragedy—for everyone involved—would be to lie without knowing it.

This discussion of axes applies to all *xy* chart types, regardless of the style you use to plot the data—whether lines, bars, or areas. In the next chapter, I'll describe how your choice of plotting style also affects the interpretation of a chart.

7

Chart Picks and Liars' Tricks

Or, Choosing Chart Types and Styles

IN **previous chapters,** I discuss various chart types and how they can be used—or abused. Here's some advice on picking the right charts for your presentation. There are also lots of pointers on knowing when someone else has deliberately picked the wrong chart for the job!

Selecting a charting format shouldn't be mainly about attractiveness of design or even personal taste. What you want to say to an audience about the data should determine the type of chart you choose to make your point. For example, you've probably grown weary of my preaching that pies are for percentages. But if that's your message, you should stick with pies. Also be aware that, having selected a chart type, important parts of your message will then be out of your control: The type of chart you select will make a specific statement about the data—regardless of how you adjust the design to suit your message. There are messages inherent in the chart types themselves, just as there are implied meanings in graphic arrows that point left, right, up, or down.

 If you don't believe me about the nuances of left/right/up/down, you might want to review my words of caution in Chapter 2.

So, making these choices isn't necessarily about picking what's pretty. It's more like grabbing the right tool for the task. This chapter should give you some guidelines so that what they think you mean comes as close as possible to what you think you told them!

Being Picky about Types and Styles

For the sake of discussion, let's distinguish between *chart types* and *plotting styles*. In this book, there are only three major chart types: pie, *xy*, and some special ones. One special type already covered is the radar chart, which has its own unique set of rules. I'll mention a few other special types in this chapter, but it's mostly about the trouble you can get into with *x* and *y*.

Plotting styles are subtler variations within a chart type. *Three-dimensional* and *exploded-slice* are two styles within the *pie* chart type. Sometimes, styles can be mixed within the same chart. It's perfectly okay, for example, to create a 3D pie that also has an exploded slice.

Selecting a charting method involves picking a chart type, then plotting styles within that type. *XY* charts give you by far the widest range of choices. This major type includes several different plotting styles—which, if you want to be truly picky, might be called *subtypes* rather than styles. Subtypes would include various kinds of bars, lines, and areas used to define and connect a series of data points. And there can be lots of variation within each of those.

LIAR'S TRICK

Liars will choose chart types and styles that hide or minimize messages that might otherwise be obvious in the data. In other words, liars will deliberately pick the *wrong* tool for the job.

Having agreed on these distinctions between type and style (okay?), let's see how your choosing among them can affect the conclusions an audience might draw from a chart's data.

Let's start by trying to plot the same set of data in a variety of ways. (This plan might not always work or even be a good idea, but let's see where it leads.) Table 7.1 shows the data in the raw, a seemingly unremarkable set of statistics, subject to no particularly alarming interpretations. It summarizes the number of ounces of liquid I drank each day in a typical work week. (This is just an example, so don't worry too much about why I bothered to keep track.)

Look first at the data series, which includes two rows of values. The top row is the days of the week. In mathematical terms, this top row

Table 7.1   *Weekday Consumption of Liquids (oz.)*

MON	TUE	WED	THU	FRI
39	42	47	60	73

is an *ordinal* set of values. Ordinal values imply some order, or sequence. The numbers 1–9 are ordinal because I can count from 1 to 9. The days of the week are ordinal because I can "count" from Monday to Sunday.

By contrast, there is no order of counting for the amounts of liquid I consumed. The numbers appear to occur randomly, although there might be some other as yet undetected pattern to the data besides their occurrence in a daily sequence.

In this example, the ordinal set of values—the days of the week—controls the order in which the other set of values can be plotted on a chart. Because the ordinal values happen to be names of things (days), they can also be used as labels to identify the data points.

Pie

That's just what I need to make a pie chart—a set of numbers and a corresponding set of labels. Each number will determine the size of a pie slice, each label a slice name. To calculate the size of a slice, divide the number by the total of all the slices, then multiply by 360 (the total number of degrees in a circle). The result will be the size of the slice in degrees:

$$\frac{NUMBER}{TOTAL} \times 360 = SLICE\ IN\ DEGREES$$

After doing this for all the numbers, I have a chart (Figure 7.1) that shows the *relative amounts* of my consumption each day.

Figure 7.1 *A pie would be a way of comparing the relative amounts I consumed each day.*

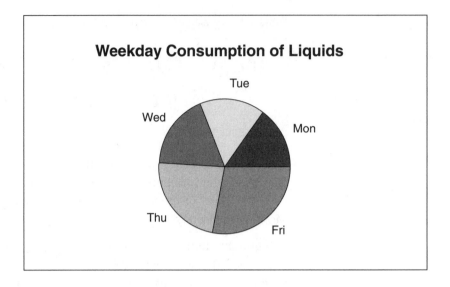

The only use I could have for such a chart would be for making comparisons among the days' results. By looking at the chart, I could draw conclusions such as, "I drank a lot more on Friday," or, "I drank almost twice as much on Friday as I did on Monday."

You shouldn't be asking questions about specific amounts on this chart, for reasons explained in Chapter 3. (You shouldn't be thinking about actual values when you are looking at a chart that shows proportions.) Nor should I, the scrupulous presenter, be telling you about the amounts as you look at the chart—and you shouldn't be inferring anything about them. Remember, from a pie chart, there is no way of graphically showing the specific amounts represented by the total pie or any of its slices. (I could label the slices with amounts, but the labels aren't features of the pie itself.)

If I showed you a pie chart, but talked about the specific amounts of the total pie or of the slices, I'd be drawing the wrong conclusions. I might be simply misguided. Or I might be lying, because you have no way of knowing these things from the chart. You might be impressed with my *discussion* of the total, which is completely unsupported by the chart. Worse, you may remember this favorable impression long after you've forgotten what I actually said.

The fact that you can't know anything about the absolute total of a pie opens the door to further trickery. A liar might advise, "Don't show the 'don't-knows.'" Because the audience has no knowledge of the total, you aren't forced to show all the data. No matter how many slices you include, they will always total 100 percent of the pie. Omitting the "don't knows" means that, rather than plotting a slice labeled All Others or Don't Know, you simply don't include it. The effect will be to make all the other slices bigger than they would be if you had plotted all the data.

In my example, it could be that the data in Table 7.1 is only a partial record of my intake. I might have forgotten or neglected to make note of everything I drank. But when you see the chart in Figure 7.1, the implicit assumption is that I'm showing you 100 percent of the actual data.

 **LIAR'S TRICK**

A pie-chart liar's trick is to omit miscellaneous slices. If the miscellaneous items make up 10 percent of the actual data sample, the net effect will be to expand the remaining 90 percent to become 100 percent of the pie. In other words, the percentage of each of the slices will be exaggerated by making each of them 10 percent bigger. (See Figure 7.2.)

Figure 7.2   *The pie on the left includes all the data in an actual sample. The Not Recorded slice has been omitted from the pie on the right, thereby distorting the percentages of all the other slices.*

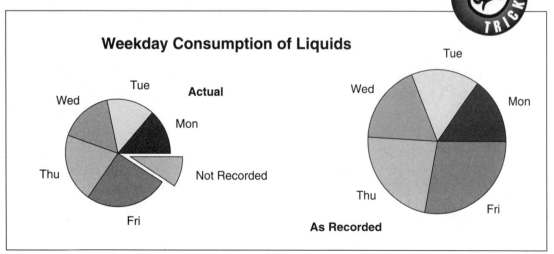

Line

Okay, so you want me to talk about specific amounts. I'll have to resort to some subtype of *xy* chart. In perhaps the most common variation of *xy* chart, if one of the sets of values is ordinal, it must be used as the *x*-axis labels. The labels are placed in sequence at major scale divisions.

The other set—the numbers—becomes the *y*-axis values. If I plot the data points and connect them with line segments, the result is the line graph in Figure 7.3.

The advantage and the disadvantage of a line graph is that it highlights any trend in the data. In fact, line graphs are used so frequently to show trends that audiences are prone to see trends in those charts even if actual trends don't exist in the data.

Returning to the example, if you tried to find a trend in my drinking habits, you might suspect that I'd break my daily record on Saturday. You might also suspect that these anonymous liquids include alcoholic beverages and that my High Friday is just that!

For the moment, let's assume that you're right and that the data is a weekly trend and at least partially descriptive of my barroom habits.

Figure 7.3 *Connecting the data "dots" creates a line graph, which might wrongly imply some kind of trend.*

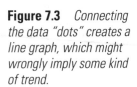

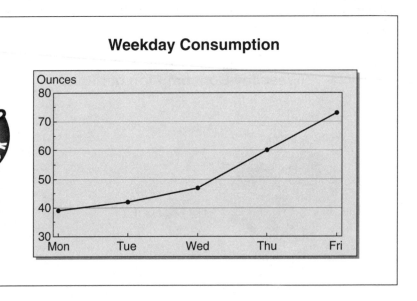

But because I'm an accomplished liar, I can make you doubt that conclusion.

Recall in Chapter 6 that two dissimilar sets of data can be plotted on the same *xy* chart by adding a second *y* axis. I'll enhance the existing chart by adding a second *y* data set—the outdoor temperatures for the week in question. I then present you with the chart in Figure 7.4. What conclusion do you now draw?

It's not that I'm a problem drinker, after all. My intake was obviously done to replenish vital bodily fluids. You can see that the peaks in my drinking coincided roughly with the high-temperature days. Not guilty! It was lemonade, and I was in the midst of a heat wave! (Notice that I've also used the tips in Chapter 6 to adjust the *y*-axis scales to my advantage.)

Trendy thinking (about numbers) is the subject of the next chapter.

In reality, the outside temperature may have had no influence whatsoever on my drinking habits. In my first try at making a chart, perhaps I made a mistake by plotting it as a line, tempting you to draw inferences about a trend that might not exist. On the second try, I have both emphasized and explained that trend by superimposing another

Figure 7.4 *Here's a really slick lie. I plotted something that really is a trend against a second y axis, implying that one set of data somehow influences the other.*

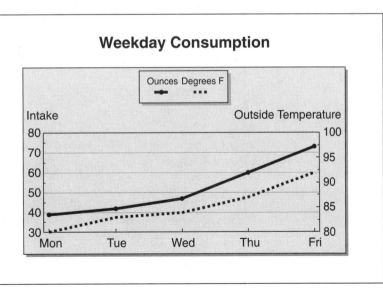

line—this one an undeniable trend. Everyone knows that the weather follows trends. And you will freely admit that the weather is beyond my personal control. Isn't my consumption of liquids, then, also out of my control?

Very neatly, I've used one trend to excuse the other, even though there may be no real cause-and-effect relationship between them.

If I look hard enough, I can find some trend in the environment that has the same pattern as my initial data plot. I can point to the value of the yen, the balance of trade payments, population growth, or a decline in purchasing power. Computer programmers even say there's a trend for trends that defy other explanations—*POM-dependency* (phase of the moon). A clever liar will superimpose such an indisputable trend on a questionable one to give the mistaken impression that the first has a controlling influence on the second.

Area

If I wanted to emphasize the *volume* of my consumption, I'd choose an area chart, which is much the same as a line graph, but with the area underneath the line shaded, or shown as a solid area (Figure 7.5).

Figure 7.5   *An area chart emphasizes both a specific volume (x times y) and a trend.*

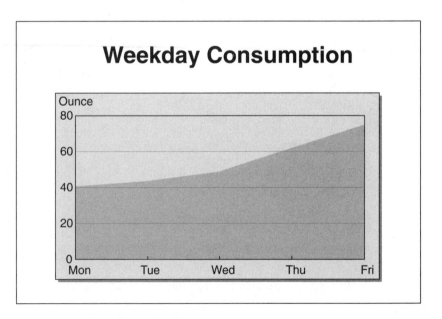

The total volume of my weekly consumption is 261 ounces, which is represented by the whole of the solid area. Unlike a pie slice, this area has a specific value, as measured against the chart scales. It can also show a trend, by virtue of the line graph that marks its upper boundary.

You can get a good approximation of the area by drawing the top of a rectangle through the mid-point of the line that defines the upper edge of the area. The size of the area is then $x \times y$, where y is the point where the top edge of the rectangle intersects the y axis.

Perhaps the best use of an area chart is for comparing the specific volumes of several sets of data. Think of it as a hybrid of both pie and xy charts—showing ratios *and* specific amounts.

It's time I came clean. Table 7.2 breaks my weekly consumption of liquids (in ounces) down into different types of beverages.

Figure 7.6 plots this data as multiple areas on the same set of xy axes.

Now, there's something you should understand about this area chart. It's data series are *stacked*, or plotted on top of one another. The base line of the first (bottom) area is the x axis ($y=0$). But the base line of the second (middle) area is the top edge (the line graph) of the first. The base line of the third area is the top edge of the second.

Table 7.2 *Breakdown of Intake by Beverage Type (oz.)*

	MON	TUE	WED	THU	FRI
COFFEE	6	6	10	6	8
SOFT DRINKS	16	18	15	20	32
FRUIT JUICE	4	0	3	10	3
WATER	13	17	19	24	30
DAILY TOTAL	39	42	47	60	73

Figure 7.6 *Multiple areas permit the volumetric comparisons of pie charts, but with the specific values of an xy chart.*

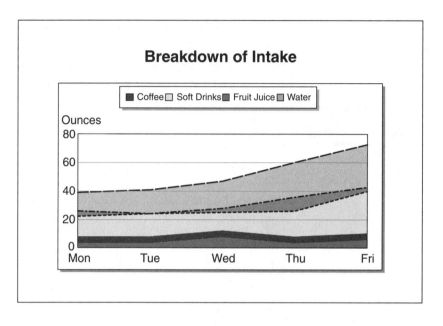

Remember the rule for using stacked areas correctly: Put the data series with the least amount of fluctuation on the bottom. Otherwise, a "bumpy" area will create distorting bumps in the areas above it.

Your audience might not understand this stacked effect. Unless you tell them otherwise, some people will assume that the base lines of all areas begin at the *x* axis. Figure 7.7 shows the two ways of interpreting an area chart if the audience doesn't know whether—or understand how—the series are stacked.

An unambiguous way of showing different areas is to plot them in 3D, as shown in Figure 7.8. In such a chart, the areas are usually *not* stacked. Here it is important to show the area with the least values (not necessarily the least variation) in front, to avoid hiding the tops of the areas in back.

A problem with 3D areas, as with some other types of dimensional plotting styles, is that it can be difficult for the audience to estimate the exact tops of the areas in relation to the *y*-axis scale. However, since the purpose of an area chart is usually to emphasize volumes, this drawback might be less important than the potential problem of interpreting the base lines of 2D stacked areas.

Figure 7.7 *If you don't know that the data series of an area chart are stacked, there are two ways to interpret how the areas are drawn.*

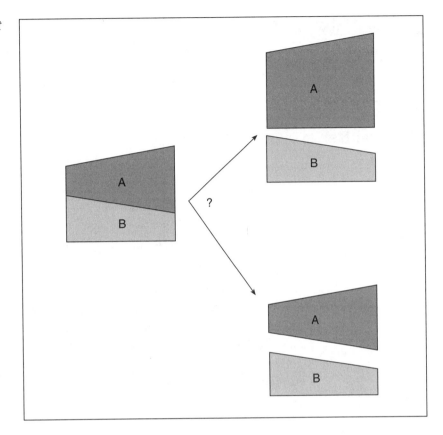

Figure 7.8 *A 3D area chart might be easier for the audience to interpret than the 2D stacked type. (Data series have been rearranged so that the areas are easier to see.)*

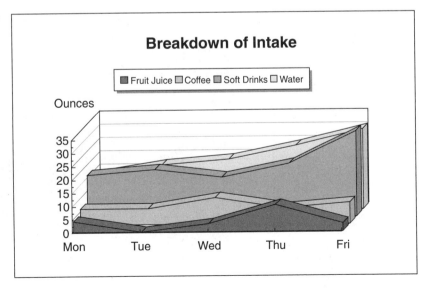

The liar's trick with areas is a fairly simple deceit: Areas simply look more *substantial* than line plots. Because an area emphasizes total volume, the audience might be less concerned with the peaks and valleys than if the data was shown simply as a line.

Vertical Bar

Plotting the first set of data as vertical bars is likely to have a very different effect on an audience. Because bars are separate, they tend to minimize trends, emphasizing the individual result at each division of the x axis. The first set of data, describing my consumption of unnamed beverages, is shown as vertical bars in Figure 7.9.

If you insist on seeing trends in bar charts, there are two variations you can try. The first is to connect the data points at the tops of the bars with a line graph. This is called the *bar-line* style. (The line need not plot the same data series as the bars, but it does, in this case.) The second is to draw line segments between the top corners of the bars. This technique is called *linked bars*. Both techniques are shown in Figure 7.10.

Figure 7.9 *Vertical bars minimize trends and emphasize the individual results at each x-axis division.*

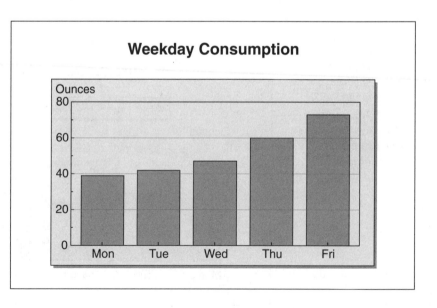

Figure 7.10 *Two techniques for high-lighting trends in bar charts are bar-line and linked-bar styles.*

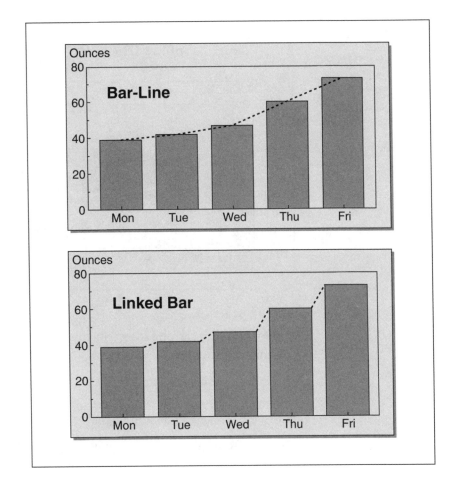

Use these bar-line or linked-bar techniques with caution. A major reason to use vertical bars is to show results from multiple entities. The *x*-axis data need not be ordinal; the labels can refer to totally separate things. For example, a bar chart could show sales results, in dollars, by salesperson. The *x*-axis data, then, might be the names Jill, John, Jake, and Jane. Aside from alphabetical order, there is no obvious way to "count" from Jill to Jane. But—regardless of their order on the *x* axis scale—the labels still make a perfectly understandable chart (Figure 7.11). This is one of the few instances in *xy* charting when the *x* axis has nothing to do with the flow of time. Furthermore, results of

Figure 7.11 *The bar chart is the only type that can be used to show separate entities (salespersons, in this case) as* x-axis labels.

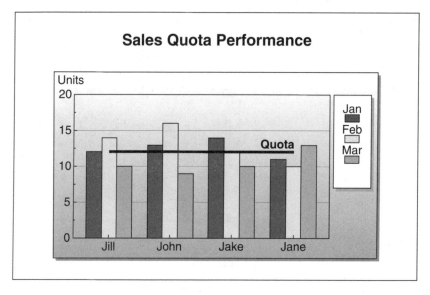

different entities can't be shown in any other *xy* chart style—you must use some type of bars.

If a chart must show multiple sets of compatible bars (all plotted against the same *y* scale), you have three different styles to choose from: clustered, stacked, and cumulative.

The data from Table 7.2, describing my consumption by day and type of beverage, is plotted in these different ways in Figures 7.12–7.14.

Clustered

In a clustered bar chart (Figure 7.12) separate bars are plotted for each data series. One bar from each series is grouped with the others, forming a cluster, at each major division on the *x* axis.

If you use the clustered style for bars, it will be fairly easy for your audience to estimate the height, or amount, of each bar. The height is obvious because all bars begin at the *x* axis base line (*y*=0).

Stacked

In the stacked bar style, bar segments for each data series are stacked like building blocks on top of one another at each *x* axis scale division

Figure 7.12
Clustered bars

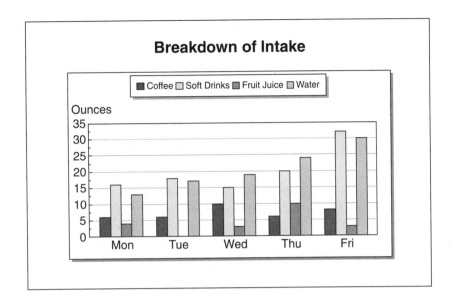

to form a single bar (Figure 7.13). Compared with clustered bars, it will be more difficult for the audience to estimate the absolute sizes of each segment (since they *don't* all start at $y=0$), but it will be easier to compare the relative sizes of segments within each bar.

Figure 7.13
Stacked bars

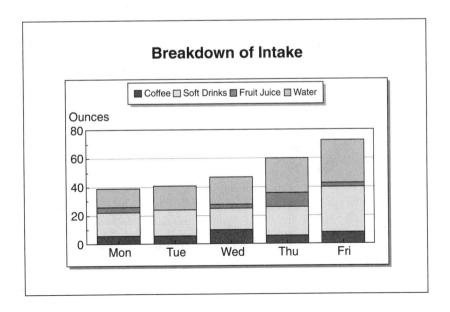

Cumulative

Bars can be *cumulative* only within a set, or data series (Figure 7.14). In a set of cumulative bars, the height of each bar represents a running total of the bars to that point. So, the value of the second data point is added to the first; and the result is plotted as the height of the second bar. The third point is added to the first two to produce the third bar, and so on to the end of the data set.

Cumulative plotting is fine for showing things that actually accumulate, such as the buyer's equity in a home or the compounded interest in an investment. Cumulative plotting is *not* appropriate—and can be a cheap trick—for amounts that aren't usually shown as running totals. Everybody's favorite numbers—sales results—would be an example. True, monthly sales accumulate during the year to produce gross revenue. However, the monthly results are usually not shown as cumulative because the audience wants to focus on how current or forecast performance compares to the results for the preceding months.

Now, here's a real note of caution: Many people confuse *stacked* and *cumulative*. Remember, stacked refers to multiple sets of data—

Figure 7.14 *Cumulative series, clustered bars. (Note the greatly expanded y-axis, which now spans 0–100.)*

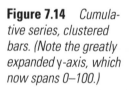

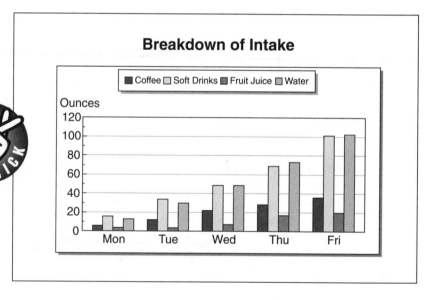

Figure 7.15   *The bars in this chart are both cumulative within each data series and stacked among series. The result is either hopeless confusion or meaningless optimism about those steadily climbing bars!*

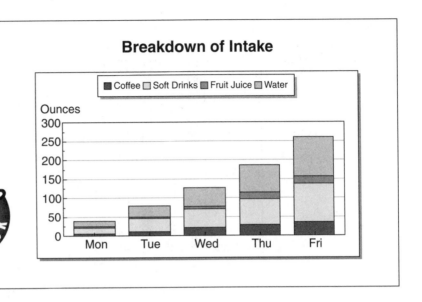

one stacked on top of the other. In charting, cumulative plotting applies *within* just one set of data—as each point is added to the preceding ones.

It's a particularly nasty liar's trick to mix stacked and cumulative techniques in the same chart, especially without making mention of the cumulative part. The example in Figure 7.15 takes mean advantage of the potential confusion of stacked and cumulative plots, showing bars that are not only deceptively tall but also steadily increasing in height— even though the noncumulative data is much less impressive.

100 Percent

A curious variation of the bar chart is *100-percent bars*. All bars have the same height, representing 100 percent of the thing the bar represents. Each bar, then, is much like its own pie chart. A set of 100-percent bars can be used instead of a multiple-pie chart, with the advantage that it will be much easier for the audience to make comparisons among bars than among pies.

Use 100-percent bars for the same reason you'd use pies—to show percentages and ratios. Although multiple pies are usually confusing

Figure 7.16 *The 100-percent bar style can be used instead of multiple pies to compare ratios.*

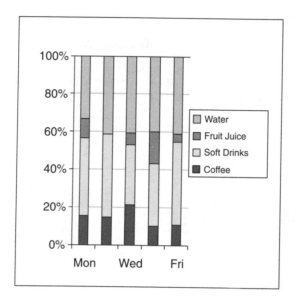

and difficult to read, the example of 100-percent bars in Figure 7.16 shows how the same objective can be achieved in a much more understandable chart.

Horizontal Bar and Gantt

To recap my Theory of X and Y discussed in Chapter 3 and 4, up-down means gain and loss, and left-right usually means the passage of time.

It's a liar's trick to show things that depend on magnitudes—such as sales—as horizontal bars. Save those lazy bars for showing time spans, such as the durations of tasks in a Gantt chart.

Paired Bars

A variation of horizontal bars—the *paired-bar* style—by its very nature permits you to transgress not one but two tenets of my Theory of X and Y. The misuse of this style shares the first violation with other

Figure 7.17 *A paired-bar chart is doubly mistaken—by using horizontal bars to show magnitude and by using leftward-going bars to show positive values.*

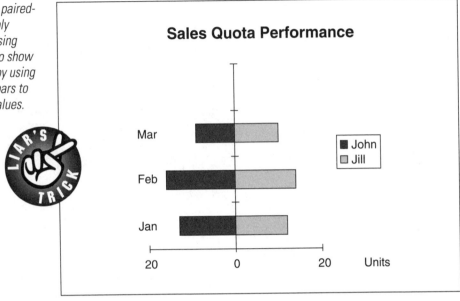

horizontal bar charts—showing magnitudes on a horizontal rather than a vertical scale. The second violation has to do with the pairing of the bars—one set grows rightward from an origin in the center, the other leftward. Now, recall that in the Cartesian coordinate system of plotting, a scale that proceeds leftward from the origin represents *negative* values. A paired bar chart, however, ignores this notion. The pairing of rightward- and leftward-going bars is simply for purposes of comparing the two data sets. So, in a paired bar chart, motion *outward*—to the right or to the left—means gain; motion inward means loss. There's an example in Figure 7.17, but, as you might expect, I'm not recommending this style to anyone.

Venn and Bubble

Circular kissing cousins in the chart world are *Venn diagrams* and *bubble charts*. If you remember your high school math teacher talking about *sets*, you're way ahead of the game here. In mathematics, a set is a group of things, usually numbers. Typically, some criteria or

Figure 7.18 *This Venn diagram shows the intersection of sets A and B.*

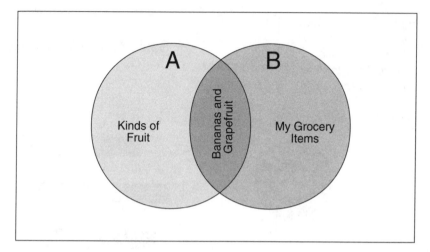

characteristics define what stuff goes in a set, such as a set of "kinds of fruit." Another set might be "items in my grocery sack." The *intersection* of two sets is what happens when you compare their respective contents. The intersection of "kinds of fruit" and "items in my sack" turns out to be a bunch of bananas and a grapefruit. Figure 7.18 shows this in a Venn diagram as two intersecting, or overlapping circles.

The *union* of sets happens when the contents of two different sets are combined, as shown in Figure 7.19. In this case, the union of "apples" and "oranges" is "pieces of fruit," shown as two circles that are entirely contained within the larger set, or category. (A set that is entirely contained within another is called a *subset*.)

Venn diagrams are for showing concepts, not quantities. The sizes of the circles can relate to the relative sizes of the groups, but more often they must be sized to permit larger circles to contain smaller circles, showing relationships between subsets and the larger sets that contain them.

Bubble charts place Venn diagrams, in effect, on *xy* chart axes. The purpose is to show how sets of things compare according to factors such as complexity (usually the *y* axis) and time (usually the *x*), as shown in Figure 7.20.

Figure 7.19   *This Venn diagram shows the union of sets A and B.*

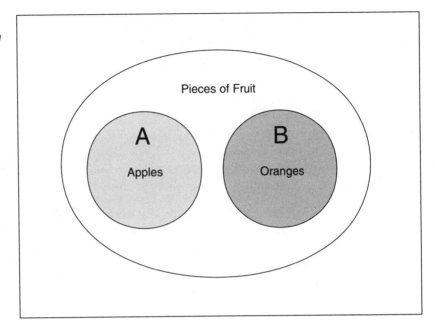

Figure 7.20   *Like Venn diagrams, bubble charts show concepts, not quantities.*

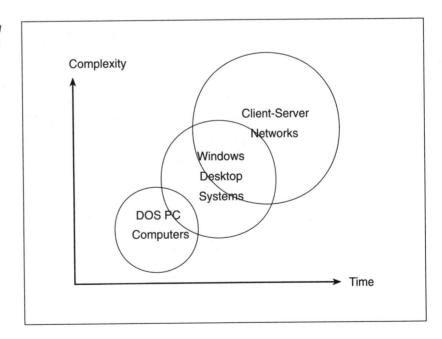

Despite this plotting scheme that looks suspiciously like an *xy* chart, bubble charts should not be used to show absolute quantities. The scales are relative, and so must be the sizes of the circles.

Mixed charts

If you try to mix several styles for different plots in the same chart, you should know that there are only a few combinations that are effective.

Refer back to Figure 7.10 to see a combination of bars with a line (a *bar-line* chart). Figure 7.21 shows points on a scatter chart with one of several possible trends traced through them as a line.

This topic of spotting trends in charts is so full of opportunities to do mischief that I devote the next chapter to pointing them out.

Figure 7.21 *Divining trends from discrete points is a primary use of the scatter-line chart.*

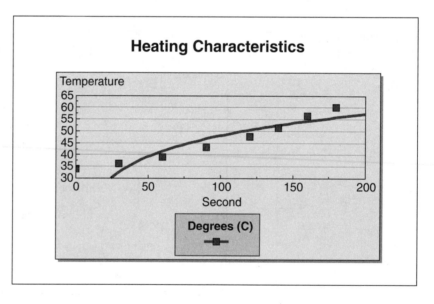

8

Trends
Or, A Conservative's Guide to Trendy Thinking

SINCE **ancient times,** seeing into the future has been an inexact science and an even more unreliable business. In Shakespeare's *Julius Caesar*, high priests dissect a dove to inspect its entrails, hoping to divine in that awful mess some clue about the Roman emperor's fate. Today, having mostly abandoned mysticism as a basic business tool, we look for signs of the times in charts. Ambitious business managers must now be enthusiastic trend-spotters, using various *trend analysis* techniques to study squiggly lines as carefully as our superstitious ancestors examined those bird intestines.

Considering the theme of this book, you can probably guess that trend analysis, like other charting techniques, can be a tool for either enlightenment or deception.

This chapter examines trends in *xy* line charts. Line charts can highlight trends in data better than any other chart type. In fact, they are used so often to show trends, that audiences are prone to view all line plots this way. They see an increasing *slope* (the techie term for line-chart *slant*) that they assume will continue upward, or a decreasing slope headed downward, or a cycle of ups and downs that must surely continue to repeat itself.

Some convention or other dictates that we show the present as a solid line, the tenuous future as dashed or dotted.

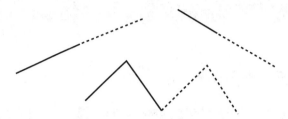

The ability to emphasize trends visually can be a powerful tool for the conscientious presenter, but it can also be just as effective when used by liars who have no qualms about exploiting our expectations. After all, a trend, if such a thing exists at all, might turn out to be something very different from the initial forecast:

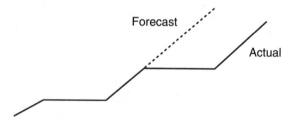

Perhaps to exploit this bias toward trendiness, line charts are commonly used by investment counselors. The intent is to show

This Slope Isn't So Slippery

In an *xy* chart, *slope* has a very precise meaning. The slope of a line for any two points is the difference between the two *y* values (called *delta y* and written (Δy), divided by the difference between the two *x* values, or

$$\Delta y / \Delta x$$

If the line slopes upward, the second *y* value will be greater than the first, and the slope will be positive. If the line slopes downward, the second *y* value will be less than the first, and the slope will be negative. The higher the number, the steeper the slope.

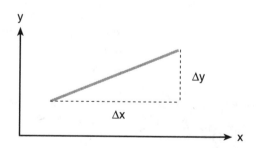

prospective investors how some pattern in the present—such as the return on an investment or the price of a stock—foreshadows future gain. However, as eager investors often learn too late, there is usually no assurance that the factors driving up today's income will continue to operate. So, the trend that looked like a trustworthy promise was, at best, an educated guess about what *might* happen.

Trend Analysis: Fact or Fiction?

Mathematicians approach trend analysis this way: Find a formula or equation that is capable of generating the data we can observe. The attempt is usually to find the simplest formula that successfully predicts future results based on numbers we have gathered in the past.

For example, faced with the data series:

$$0, 2, 4, 6, 8$$

The mathematician might derive a formula something like:

$$\text{NEXT NUMBER} = \text{PREVIOUS NUMBER} + 2$$

Having found such a formula, a mathematician might then use it to predict future results:

$$0, 2, 4, 6, 8, 10, 12, 14, 16$$

Viewed as a chart, the trend seems perfectly obvious.

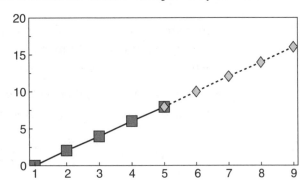

But the pattern in the actual data might not be so simple. When the future is now, so to speak, the actual data might look like this:

0, 2, 4, 6, 8, 3, 5, 7, 9, 11

There is a pattern here as well, but it isn't apparent from inspecting the first four numbers. The formula for this series might be

```
START AT 0
NEXT NUMBER = PREVIOUS NUMBER + 2
DO FOUR TIMES
START AT 3
NEXT NUMBER = PREVIOUS NUMBER + 2
DO FOUR TIMES
```

As with the straight sloping line in the example above, it's much easier to spot the trend in a chart:

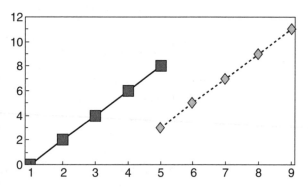

The mathematician's formula amounts to a guess about the process that is generating the numbers. The pros have a word for extending the past into the future—*extrapolation.*

A related technique called *interpolation* attempts not to project forward, but to fill in the gaps in a set of historical data.

0, 2, ?, ?, ?, 10, 12, 14, 16

So, here's one guess at the missing numbers:

0, 2, **4**, **6**, **8**, 10, 12, 14, 16

Again, it looks so easy when you describe the problem in a chart:

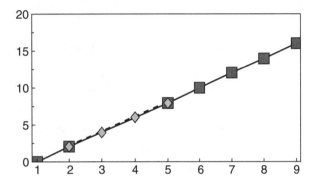

In practice, extrapolation and interpolation work well if the underlying formulas accurately describe the real forces that are shaping the events we observe. This approach works best for simple formulas—such as the compound interest on a certificate of deposit (provided that the depositor makes no unpredictable withdrawals!). Extrapolation and interpolation are also pretty reliable at describing the effects of natural laws, such as the flight of a ball through the air. Such predictable processes depend on just a few basic factors—the force and direction of the pitcher's throw, the force of gravity, the mass of the ball, and wind resistance. But when you leave the ballfield and venture into a complex field such as economics, the forces are many and largely unpredictable- at least by simple formulas.

The Lessons of Oversimplification

However complex or unpredictable the real forces may be, a line chart that describes the results can retain the *appearance* of simplicity, along with the false assurance that today's patterns will be somehow repeated in the future. It's a universal human tendency to try to see patterns, which in most aspects of daily life is a mark of intelligence. But as you are perhaps beginning to suspect, this tendency can get you into serious trouble, chartwise.

Let's take a closer look at some specific types of trends you might see in chart data. I'll describe what these trends can mean, as well as how they can be used to distort rather than to accurately describe reality.

Beware of Your Average Chart

There are no pop quizzes in this book, but some of the stuff from Chapter 1 should come in handy right now. Recall from the example of the fibbing golfers that an arithmetic *average* is the sum of the scores of previous games, divided by the number of games. Just to make things interesting, let's assume that the same golfer shot an unusually high 92 last time out:

$$\frac{79 + 81 + 78 + 80 + 76 + 92 \text{ STROKES}}{6 \text{ GAMES}} = 81 \text{ AVERAGE STROKES PER GAME}$$

In a line chart, the average is one type of trend, as shown in Figure 8.1. The individual scores form the peaks and valleys of a line plot, and the

Figure 8.1 *The golfer's average score is shown here as a horizontal line that neatly cuts through the peaks and valleys of high and low scores. Also shown are the mean and the median— very different calculations that produce similar-looking trend lines.*

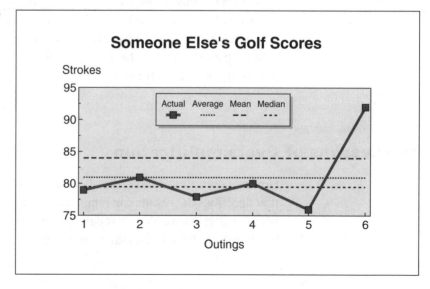

average is shown as a straight horizontal line that cuts neatly through them, leaving a symmetrical pattern of bumps on either side.

He Shoots a Mean Game

To unsuspecting audience members, the golfer's *mean* score means much the same as the average. To get the mean, you would take the average of only two values, representing a high and a low, by adding them and dividing by 2.

In the case our favorite golfer, the high was that pathetic 92 and the low was 76. The mean is

$$\frac{92 + 76}{2} = 84$$

For the charted result, refer back to Figure 8.1. Remember that the mean can be a better trend indicator if the values are wide-ranging, as they are here. For this golfer, the 84 mean might be a better predictor of his next game than his average of 81, which might be a better forecast of performance over a longer period of time.

 The mean is sometimes called the *mid-point,* but that's too easy to confuse with *median,* which is something different.

Know What They Median

In a group of data values, the *median* is not an average, but one of the values that seems "normal." The median value in a set is located exactly in the middle—not in the middle of the range of values but in the middle of the number of items. So, there are the same number of values above the median as are below it. If you had a set of five scores, there will be two scores above the median and two scores below it. If you had an even number of data values, you would take the mean of the *middle two* as the median. For example, in a set of four values, you would take the mean of the second and the third values, so that there are two values above the mean value, and two below it.

Here again is the data for our golfer:

$$79, 81, 78, 80, 76, 92$$

There are six scores-an even number of items. The middle two scores are 79 and 80. (There are three scores below 80 and there are three above 79.) The median is

$$\frac{79 + 80}{2} = 79.5$$

 Look back at the chart in Figure 8.1 to see how this method of calculating a typical score might distort the truth. *Using the median has the effect of ignoring exceptional highs and lows.*

When in Doubt, Keep Moving

Recall that one of the golfer's last resorts might be a *moving average,* which would always include only the few most recent set of games-say, the last four. As each new game is included in the average, the oldest game drops off so that there are always just four games included. A chart of long-term performance, then, would show not only the actual scores, but also a series of four-game averages (see Figure 8.2).

 Liars can use *average, mean,* and *median* interchangeably in chart descriptions, picking deliberately among these techniques to show the one that gives the most favorable result. A similar but perhaps subtler trick is to show a *moving average,* which might include only recent, favorable data, instead of using a longer-term *average*—and neglect to point out the "moving" part!

Figure 8.2   *The smoother line is the player's four-game moving average, which some people might regard as a better description of overall progress.*

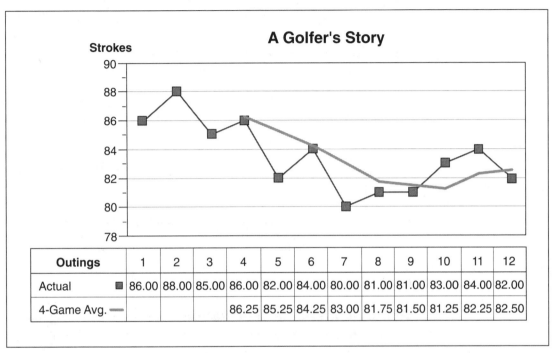

A Golfer's Story

Outings		1	2	3	4	5	6	7	8	9	10	11	12
Actual	■	86.00	88.00	85.00	86.00	82.00	84.00	80.00	81.00	81.00	83.00	84.00	82.00
4-Game Avg. ━					86.25	85.25	84.25	83.00	81.75	81.50	81.25	82.25	82.50

When Is Deviation the Right Road?

Statisticians are naughty mathematicians who, when kept after school, decided that they *liked* copying the same sentence a thousand times. Seriously, though, some of my best friends are statistics freaks— including many of you who were wise enough to buy this book!

More to the point, statisticians are world-class trend-spotters. However, like any group of professionals, they have developed their own jargon, which can be downright intimidating to the average (or mean) chart-reader.

When speaking of the *mean* (as in the foregoing example), statisticians make the following distinctions, which may or may not have anything to do with what you want to know about a chart:

Standard deviation is the degree to which values in a series of numbers deviate from the mean. The standard deviation is usually shown as plus or minus the difference between the high and the mean, and the low and the mean. For example, if the mean is 1, the high 3, and the low −1, the standard deviation is ±2.

Population standard deviation is just a way of saying that all items in the data series were taken into account in calculating the standard deviation. The population standard deviation is used when the data series, or list of numbers, is small enough to be examined in its entirety.

Sample standard deviation indicates that only a typical selection, or sample, of data items was used to calculate the standard deviation. The sample standard deviation is used when the data series is very large, including perhaps millions of items.

In a bar chart, the standard deviation can be shown as *error bars*, indicating a typical range of values. More often, the presence of error bars indicates a degree of uncertainty: The chartmaker is telling you that the actual value might lie anywhere within the upper and lower bounds of the error bar.

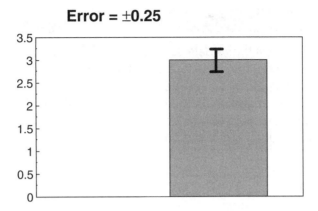

Error = ±0.25

Liars are not usually honest enough to even think of using error bars, much less make the distinction between *population* and *sample* standard deviation. Scientists, who are supposed to tell it like it is, often show experimental results with error bars. The deviation might indicate, for example, limits on the accuracy of measuring devices.

How Can Regression Advance Your Career?

In the context of charting, *regression* doesn't have its psychoanalytical meaning—to return to a more primitive behavior—unless, of course, a chartmaker is monkeying with the truth!

Regression is another trick of the statistician's trade for finding patterns in data. By trying different regression formulas, the statistician can fit smooth lines neatly through seemingly erratic sets of data points. The assumption is the better the fit, the more likely that the formula actually describes the process that is generating the data.

The process of applying this technique is called *regression analysis*. For math mavens, the algebra is in the sidebar, "Regression: The Gory Details." For the rest of us, it might be sufficient to know that regression calculations are built into business graphics and spreadsheet software.

Regression: The Gory Details

To generate a regression line, the data set must contain at least three data points. In practice, there should be many more to assure that any curve fit is meaningful. Different types of regression, defined by different equations, can be applied to the data to see which of the resulting lines produces the best fit through the points. The best fit is assumed to be the best model for predicting future results. Common regression types are linear, exponential, logarithmic, and power. Here are the formulas for each type:

Linear $\qquad y = a + bx$

Exponential $\qquad y = ae^{bx}$

Logarithmic $\qquad y = a + b\,(l_n x)$ $\qquad (l_n x)$ is the **natural logarithm** of x

Power $\qquad y = ax^b,$ where $a > 0$

In these equations, the dependent variable is y (the golf score in the example). There are three independent variables: $x, a,$ and b (age, height, and weight).

Figure 8.3 shows how two types of regression curves—linear and logarithmic—can be fitted to the same set of data points.

You can use regression analysis to analyze how a *single dependent variable* is affected by the values of one or more *independent variables*—for example, how a golfer's score (the dependent variable) is affected by such factors as the person's age, height, and weight (independent variables). The formula that defines the link between the dependent and independent variables is called a *correlation*. You could allocate the relative importance—or *weight*—of each of these three factors based on a set of actual performance data. You could then use the results to predict the score of a different golfer for whom you know the contributing factors but have no actual scores.

Figure 8.3 *Two types of regression analysis have been done on the same set of data points: linear and logarithmic. The equation that produces the best fit is assumed to be the best model of the data. The linear regression emphasizes the downward trend. The logarithmic implies that the downward trend will level off eventually.*

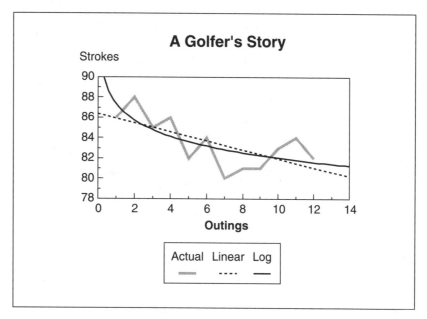

In reality, there may be no valid correlation between a golfer's scores and his or her age, height, and weight. But if you get a regression line to fit the data points, you will be strongly tempted to think that a correlation exists!

Regression analysis can be tempting for sincere forecasters and liars alike because it produces smooth line plots and curves that can appear to fit the data neatly. To the extent that there is a good fit, the underlying regression formulas would seem to describe the real forces that are shaping the data. The pitfall, as with looking for other kinds of trends in charts, is the potential to oversimplify, especially if the sample is small or if some data points have to be excluded from the plot just because they don't fit.

The liar's trick with regression analysis is simply to exclude data points that don't fit some impressive curve. In the case of experimental data, for example, the excluded points are judged *anomalous*—statistics jargon for "I don't know what to make of these." Excluding anomalous data can be legitimate, though, as when making a large number of observations there will be some errors from inaccurate measurements or mistakes in transcribing the data.

A major hazard of computer-generated trend-lines—as produced by regression analysis—is the temptation to rely on them simply because an electronic brain did the calculating. Before you jump to such a conclusion, remember—computers can lie just as readily as people can. Maybe even more so, since computers are literally without shame. Ethics and moral judgment aren't embedded in their programming. The subject of the next chapter is developing some healthy skepticism about chart data that is generated by computer.

Tables and Spreadsheets
Or, The Socialite's Guide to Setting a Nice Table

As **every dinner-party impresario knows,** much of your success can depend on how you set a table. How you set (up) a table in a report or presentation can determine whether it will inform or confuse your audience. Furthermore, if you use a table as a worksheet for chart data, you will find that the arrangement of its columns and rows strongly affects your ability to generate meaningful graphs from the data.

The stuff in those graphic tables is particularly irresistible to avid number crunchers. To these information gluttons, a pretty sales curve is always suspicious; instead, they demand,

Their preference for seeing data laid out in tables is a cultural bias just as powerful as those notions of up, down, left, and right (which I expose shamelessly in Chapter 2). This bias comes not from geographic or ethnic background, but from the time-honored traditions of the accounting profession. I'll explore that notion in some depth in this chapter.

There's No Pattern on This Sheet!

Tables are not only boring, but also obscure. It's difficult to see patterns in the data when you are faced with a dizzying array of numbers. One of the reasons sales charts are curvy is for clarity!

Liars might prefer tables instead of charts if they are trying to obscure patterns in the data. If the array of numbers in a table is particularly dense and difficult to read, the presenter either is trying to hide the truth or is "graphically impaired" (more to be pitied than scorned).

As a general rule, you should do everything you can to use a graph instead of a table to present numeric results. Tables are better as worksheets than as formats for presentation. But, if your guests demand priority seating, I offer here some information etiquette for setting a nice table.

Tables and Sheets—You've Got It Covered

As I've said, notions about tables are rooted in financial accounting. Perhaps it's the close relationship between tables and spreadsheets that makes their arrangement by columns and rows seem so comfortable to some people. For accounting types, the spreadsheet—a kind of ledger—is the basic format for their working documents, or *workpapers*. (In the jargon of the trade, it's just a *sheet*.)

Spreadsheet Style—from Accounting 101

As a quick guide for the rest of us, the basic layout of an accounting-style spreadsheet is shown in Figure 9.1. The first row holds column headings. The first column holds row headings. The last column holds totals of each row. The last row holds totals of each column. The rest of the positions in the table hold numeric data. At the intersection of the last column and the last row is the *grand total*, which can be produced by summing either set of totals. Strictly speaking, all of the other totals

Figure 9.1 *In a conventional accounting spreadsheet, being able to generate the grand total by cross-footing—summing either down the last column or across the last row—provides a double-check on its accuracy.*

Third Quarter Office Expenses

	July	August	September	Totals
Wages	5,600	5,600	5,600	16,800
Supplies	436	327	408	1,171
Rent	1,200	1,200	1,200	3,600
Utilities	659	632	749	2,040
Totals	$ 7,895	$ 7,759	$ 7,957	$ 23,611

in the sheet are *subtotals* of the grand total. By accounting convention, subtotals should be highlighted by a single underline, the grand total by a double underline.

The accounting term for summing a series of numbers is *footing*. The ability to generate the grand total in two different directions is called *cross-footing*. Particularly when spreadsheets were done manually (that is, with sums done mentally, on an abacus, or with a calculator), cross-footing provided a quick check on the accuracy of the grand total: If you get different answers, at least one of the totals in the spreadsheet must be incorrect. If you get the same answer each way, the grand total is *very probably* correct. (You could have two or more errors, one compensating for another, but that's unlikely.)

Nowadays, a spreadsheet is an electronic worksheet generated by computer. Because the computer does all the calculating, there isn't as much need to prove the accuracy of the results by cross-footing.

Cross-footing is only one basic type of verification on the accuracy of a spreadsheet. At the conclusion of this chapter, I'll give you my Eight Great Steps to assuring the reliability of spreadsheet results.

Even though cross-footing is becoming rare, financially savvy people still expect to see totals in one of two places—at the right end of rows or at the bottom of columns.

Since a computer-generated sheet is much more fluid—and therefore changeable—than ink on paper, it can be much more convenient to show totals in the *top* row, as shown in Figure 9.2. In this arrangement, a kind of "perpetual ledger" can be created. In the example, the sheet is a check register. As each new entry for a check or a deposit is made in a new row at the bottom, the sheet grows longer. But the totals don't have to move. The updated numbers just overwrite the old ones in the top row. In a manual sheet, this would be messy and impractical. You'd have to use an eraser on the totals each time a new entry was made, quickly wearing holes in the paper. On paper, when you're at the bottom of a sheet, it's more natural to simply start a new one, carrying the balance forward.

Figure 9.2   *The electronic age makes alternate sheet layouts possible, but people have the entrenched notion that totals belong at the right side or at the bottom of a table.*

Date	Reference	Checks	Deposits	Balance
Totals		**$244.92**	**$250.00**	**$1,005.08**
			Start Bal. >	$1,000.00
01/03/95	721	$15.95		$984.05
01/05/95	722	$64.35		$919.70
01/06/95	723	$21.50	$250.00	$1,148.20
01/06/95	724	$40.00		$1,108.20
01/06/95	725	$103.12		$1,005.08

LIAR'S TRICK

Presenting a table that has an unconventional layout, such as showing the totals at the top, can be disorienting to an audience. As you know from seeing liar's tricks with other chart formats, anything that runs counter to audience expectations can become another trick of the liar's trade.

Mixed Data Types—Another Cute Trick

If a liar rearranges the totals in a table to confuse you, the trick will be fairly obvious. Frankly, the result is so odd-looking and unusual that most people wouldn't even think of trying it—much less tricking you with it.

But another trick of table layout is more subtle—and much more common. It has to do with mixing types or categories of data within the same column of a sheet or table.

Tables that are the easiest to read and understand follow the same rules used in constructing both traditional accounting spreadsheets and their modern-day variant—the computer *database* table. There's an example of a database in Figure 9.3.

Figure 9.3   *Because it is constructed according to strict rules that prohibit the mixing of data types within columns, a computer database table is very easy to read and understand.*

Monthly Professional Billings

Name	Job Skill	Area	Phone	Hourly Rate	Time	Gross Billing
Arnold, B.	Tax	818	555-9284	$200		$0
Delon, N.	Audit	818	555-2176	$200		$0
Enberg, A.	Audit	818	555-3876	$100	2	$200
Garcia, M.	Accounts	415	555-2019	$100	3.5	$350
Hofstra, G.	Litigation	213	555-3917	$200		$0
Kenworth, W.	Consult	310	555-1213	$100		$0
McMasters, O.	Tax	213	555-7856	$150		$0
Nanterre, B.	Accounts	213	555-2398	$100		$0
Ono, W.	Audit	213	555-1111	$200		$0
Otto, Y.	Audit	818	555-3331	$200	0.5	$100
Pulaski, E.	Accounts	213	555-6549	$150	1.2	$180
Tan, A.	Litigation	310	555-0432	$350		$0
Tso, J.	Tax	310	555-2857	$100		$0

Each entry in a database table is called a *field*. Each column heading in the first row is a *field name*. The field name describes one and only one type of entry, which applies to all the fields in that column. All the fields that describe one *entity*, such as a person, comprise one row. A row that describes an entity is called a *record*.

Look back at Figure 9.1, the traditional accountant's spreadsheet. Except for that last row of totals—which often holds numbers that are used only for purposes of cross-footing—the sheet follows the rules exactly for construction of a database table.

So, both the accountant's spreadsheet and the computer database table show data according to rules that fit audience expectations, including the requirement to keep different categories of data in separate columns.

Intermixing data categories in the same column of a table can be another disorienting trick of the accomplished liar.

A table that violates these rules is shown in Figure 9.4. You can see right away that it's a kind of apples-and-oranges layout. This graphic is actually not a table but a *list* of dissimilar (but perhaps related) items, presented in tabular form. A good table is useful in part because it invites *comparisons* among similar entities—usually, one entity in each row. But in the liar's example, no meaningful comparison is possible among the rows. The trick is precisely to *prevent* the audience from making comparisons.

In this example, comparisons are difficult within a data category (such as Test Scores) because each day's score would have to be shown on a different slide or page of the presentation. Notice how much easier it is to compare the daily scores when they are presented in the same table (see Figure 9.5).

Figure 9.4 *This table violates the rule about showing only one data type in each column.*

Student Testing Environment

Day	Monday
No. of Students	32
Room Temperature	75.6
Test Score (Avg.)	82.0

Figure 9.5 *This table keeps data categories consistent within each column, making it possible to make comparisons among rows.*

Student Testing Environment

Day	Students	Temp.	Score
Monday	32	75.6	82.0
Tuesday	28	78	81.5
Wenesday	29	73.4	80.9

Information Etiquette for Table-Setters

Given the rules for constructing tables that people can readily understand, here are some tips on using tables (and electronic spreadsheets, if you must) in presentations.

Use tables to make your explanations more concise.

In a report narrative or a speech, you can avoid lengthy explanations of key relationships by summarizing them in a table. You can then use fewer words to simply highlight specific entries in the table or to draw overall conclusions about relationships among the entities. Be guided by the fact that businesspeople are always in a hurry—they won't be reading your report or attending your presentation for pleasure.

Don't present data in a table or spreadsheet if a graph can show the same idea.

Raw data is seldom as interesting as a chart, which can make a trend or relationship understandable at a glance. If the source data must be shown, consider presenting the numbers in a table beneath the graph:

	Aardvark	Aero	Acme	Able	Archer	Argo	Ace
Last Year ■	93	71	99	94	121	105	60
This Year □	124	83	162	95	143	231	78

Notice that the *data table* shown above contains not only the graph data but also a *legend,* or data series key, with color codes for each row. Legends are discussed in the next chapter.

When covering dissimilar things, summarize them as topics in brief lists, not in tables.

Use bullets when presenting topics and key points, especially if you are listing topics that will follow as subtitled sections of your material. Use a numbered list instead if you a presenting a sequence of steps. Professional presenters call this technique *signposting*, because it can guide the audience through the outline structure of your presentation.

Limit the number of columns in a table.

The controlling factors on the number of columns in a table are page or screen width, margins, and text size. So, you can usually increase the size and therefore the legibility of the text if you use fewer columns. You might productively eliminate a column used for explanations by placing the information instead in a footnoted list at the bottom of the table.

In a printed report, try to keep a table on a single page, close to its explanatory text.

A page break in the middle of a table makes it that much more difficult for your readers to follow, especially if they will be making comparisons among the rows. Instead of breaking a table, consider placing it on a separate page, even if it must be located farther away from its explanatory text. Give the table an exhibit or figure number, and include that reference in the text.

Justifiable Justification—Here Are the Rules!

The rules for justifying, or aligning, table entries are pretty standard. If you follow them, your tables will not only be easy to read, they'll also look neat, which always gets you extra points in my book. The rules are:

- Align alphabetic text, such as labels, left:

Acme	
Arrow	

- Align numeric values right, usually on the rightmost digit:

Acme	253
Arrow	4,841

- If numeric values are decimals, align them on the decimal points:

Acme	253.07
Arrow	4,841.372

 You needn't be concerned about aligning digits on their decimal points if you use the same number of decimal places for each entry in a column and align them all on the right.

- Optionally, center the headings in their columns:

Company	Index
Acme	253.07
Arrow	4,841.372

Publishing Spreadsheets, If You Must

All the tips about tables also apply to computer-generated spreadsheets—which, for the most part, are just big tables. You should try to avoid including spreadsheets in your reports because they are, well, just plain boring. Furthermore, your purpose should be to summarize information, and spreadsheets typically have too much detail. Again, you might communicate your message better in a graph that uses just a few critical rows from your spreadsheet as data series.

But, if you must, you must. Sometimes they insist on seeing the numbers.

The rest of this chapter is not so much about laying out the numbers visible on sheets but about the calculations underlying those numbers. It's crucial to the truth of your tables—as well as to the veracity of the charts you make from them—that you get a passing grade in spreadsheet logic. If you don't understand how the numbers are being generated, you leave yourself open to all kinds of unintentional errors—and to intentional distortions made by liars who aced the course!

The spreadsheet examples in this chapter refer to Microsoft Excel, but the concepts apply equally well to Lotus 1-2-3 and other spreadsheet programs.

Recall from Chapter 1 that a liar's first step in creating deceptive charts can be to manipulate the data from which the charts are generated. Well, as Big Mama says in *Cat on a Hot Tin Roof,* the biggest problems in life are usually under the sheets!

Finding Errors in Spreadsheets

Computers are speedy and meticulous calculators, but their programs know nothing about the larger business problems they are solving. They therefore cannot catch many kinds of human error, particularly errors in the way problems are described in math formulas.

Spreadsheets: The Crash Course

You needn't skip the rest of the stuff in this chapter just because you haven't yet experienced the joys of electronic worksheets. Although rocket scientists rely on spreadsheet technology, you don't have to be one to use it.

The columns of a sheet are marked with letters, starting with **A**, across the top:

Rows are identified by number, starting with **1** and running down the left side of the sheet:

The intersection of each column and row forms a small box. Each box is a storage location for once piece of stuff, such as a number or text. Each box is called a *cell*, and each cell has an *address*. The address of a cell is the letter of its column and the number of its row, with the column always coming first. So the address of the top left cell in the sheet is **A1**. (I don't know about you, but I find the term *cell* a bit chilly. If it makes you more comfortable, call it a "cubbyhole.")

Each cell has its own address, and no two cells have the same one. If you tell the program that you want the stuff you stored in cell D5, it will quickly fetch the contents of the box at the intersection of column D and row 5.

Picking Your Favorite Cell Block

It's often necessary to refer to a block of cells, such as the numbers in a data series for a chart. In spreadsheet jargon, a *range* is a rectangular block of cells defined by its first cell and its last cell. The first cell is in the upper-left corner of the range and the last cell is in the lower-right corner of the range. Figure 9.6 shows the range **B2:D4**. When you specify a range, you are selecting the contents of its cells. The colon (**:**) means "and all cells through and including." (Some spreadsheet programs use a different notation: **B2..D4**.)

Figure 9.6   *Here's the block of cells (or spreadsheet range) B2:D4, defined by its top left corner (cell B2), and containing all the cells through and including its bottom right corner (cell D4).*

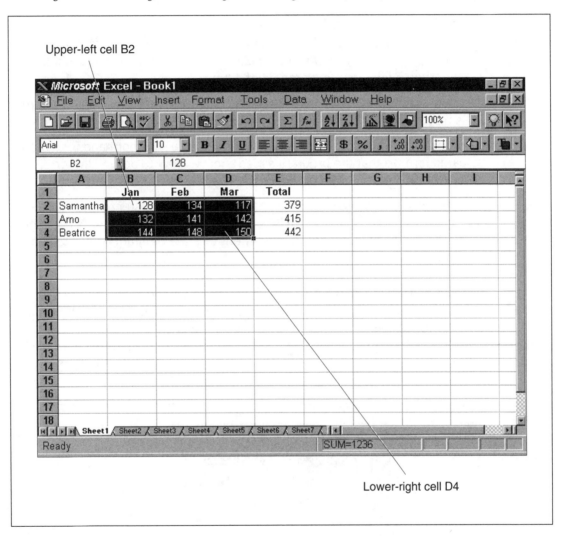

Upper-left cell B2

Lower-right cell D4

Name Your Range!

There's a much more understandable way of referring to ranges—you can give them names. Of course, this is especially handy if the name you pick describes the data that's stored there, such as **SALES**.   *(Continued)*

Spreadsheets: The Crash Course *(Continued)*

Letting the Computer Do Your Math

Instead of holding a piece of data, such as a number or some text (called a *label*), a cell can hold a little math problem, or *formula*. A formula is a calculation that the computer will do, saving you the effort. Here are the elements of a typical formula:

■ **Equal sign**   A formula usually begins with an equal sign (**=**). It tells the program, "Do this math!" (In some programs, a formula must begin instead with **@** or **+**.)

■ **Function name**   A function name—**SUM**, for example—is shorthand for a more complicated formula that the program already knows how to do. **=SUM** (or **@SUM**) adds a series of numbers; **=AVERAGE** (or **@AVG**) averages them.

■ **Argument**   Range addresses or range names enclosed in parentheses form an *argument*. With functions such as **SUM** and **AVERAGE**, the arguments describe a block of cells that holds numbers. An argument tells the program where to find numbers to give to a function.

HERE'S A FORMULA:
=SUM(B2:B7)

This formula simply means, "Sum the numbers in the range from cell **B2** through and including **B7**." The program does the work and puts the answer in the cell that contains the formula. The use of this formula in a sheet is shown in Figure 9.7.

Figure 9.7   *Here's how a formula that uses SUM might be used to add a row of numbers in a spreadsheet.*

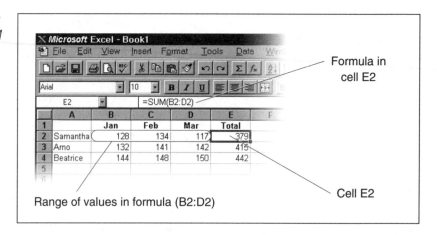

Formula in cell E2

Range of values in formula (B2:D2)

Cell E2

That's it! That's all even rocket scientists need to know about how spreadsheets work.

Spreadsheet errors can be intentional—deliberate efforts by dedicated liars—or unintentional—mistakes by spreadsheet makers whose logic is flawed. *By far the majority of errors are not committed on purpose.* I don't mean to say that there aren't many liars, there are just far more dumb mistakes! And, in a way, that's a scarier situation, because a lot of people out there in computerland are generating misleading data with the best of intentions.

For the rest of this chapter, I'll explore some common sources of spreadsheet error which can affect both tables and charts. I'll describe two basic mistakes: 1) picking the wrong formula, and 2) using flawed logic. There are many variations on these, but you'll get the idea. Then, I'll conclude by giving you my Eight Great Steps you can take to check a sheet for all types of errors.

Working with Decimal Values

Formulas that use decimal values can produce errors that are difficult to spot, even for experienced users of spreadsheet software. Getting a reliable answer involves picking the right function as well as understanding its effect on the rest of the math in a spreadsheet.

Decimal values are problematic because a spreadsheet cell can only hold a specific number of digits. This fact, in itself, is a potential source of error, since some decimals could go on forever. For example, 10 divided by 3 equals

$$3.33333333333333333333333333333333333333333333333333333\ldots$$

That ellipsis (...) means "and so on," so what you see is just an approximation, despite all those digits.

Most spreadsheet programs provide a variety of ready-made functions you can use to manage decimal values that have a lot of digits. These functions all work differently, so picking the right one is essential to getting correct results. Two particularly troublesome functions—*rounding* and *truncation*—give similar results that can create big differences.

Rounding

The function ROUND does the seemingly simple job of rounding off a long decimal value to a shorter, more manageable one. In so doing, it can adjust the last digit of the result so that it better approximates the original, longer number.

A rounding formula typically takes two arguments—the decimal value you want rounded and the number of decimal places you want in the result:

$$=ROUND(B2,2)$$

In this example, if cell B2 held the value 1654.9876, the result would be 1654.99 (with the last digit rounded up). ROUND can be the most accurate way of dealing with unruly decimals—provided that the result can also be a decimal.

The second argument you feed to ROUND can be zero or a negative number. If the second argument is zero, the program will round the result up or down to the nearest integer. For example, an argument of 0 would round 34.9 to 35. If the second argument is negative (less than zero), the program will round the number to the left of the decimal point at the position indicated by the number. So an argument of −1 would round 34.9 to 30.

Truncation

A function that works very much like ROUND is TRUNC, which truncates, or just chops off, some or all of the decimal digits. Like

ROUND, TRUNC takes two arguments—the number you want to truncate and the number of decimal places you want in the result:

$$=TRUNC(B2,1)$$

In the formula above, if B2 held the value 1654.9876, the result would be 1654.9. If you omit the second argument, it is assumed to be zero, in which case the program will throw away the entire decimal portion, leaving an integer (whole number with no decimal places) as the result.

Sources of Error from Approximations

Serious errors from ROUND and TRUNC stem from the fact that they produce *approximations* as results. The differences in a single answer might be small, maybe insignificant. The errors grow larger, though, when the results of these functions are then used by other functions, especially in multiplication. If a small error is multiplied many times, it can become big enough to matter.

LIAR'S TRICK

In a famous case of "Truncation Abuse," a bank employee noticed that a computer program used truncation instead of rounding to calculate interest on savings account deposits. The employee modified the program to send the discarded part of the truncation to his own account, eventually accumulating a considerable sum. (Before you get any ideas, the Feds are wise to this one!)

Other Similar Functions

Spreadsheet programs have many, many built-in functions—too many to cover in this modest book. But ROUND and TRUNC are representative of lots of other groups of functions that do similar things but can produce significantly different results. For example, you might pick the function INTRATE to calculate the rate of return on an investment. INTRATE *might* be correct if the investment is a bank certificate of deposit. But if the investment is a bond, you should use YIELD, and if it is a U.S. Treasury bond, TBILLYIELD. These are functions which have formulas specifically designed for those types of investment.

Calculating investment return is particularly prone to errors (or intentional misstatement) due to repeated multiplication because the return will be *compounded,* or reinvested periodically. So, the result of even a small margin of error will be literally compounded, growing ever larger over the period of investment. If a liar were trying to sell you an investment, he or she might try to overstate the return, which can be done easily with such "funny" math.

Running in Circles

Another source of spreadsheet error that can be particularly troublesome is the *circular reference.* A circular reference is the result of a formula that refers to itself.

Direct Circular References Are Easy to Spot

One type of circular reference is *direct.* This happens when you enter the address of the current cell in the formula that it holds. For example, cell **A1** might hold this formula:

$$= A1 + 1$$

Unless it appears anywhere *but* in cell **A1**, this formula describes a logical impossibility. When you enter the formula there, the program replies, "Cannot resolve circular references." However, the program will go ahead and insert the value 0 (zero) in the cell. At this point, your sheet contains a logical error, which isn't obvious from that innocent-looking zero.

Indirect Circular References Can Be Hidden

The indirect kind of circular reference looks less like a mistake, and is therefore both easier to make and much more difficult to trace. For example, cell **A1** might hold the formula

$$= B1 + 1$$

This, in itself, is perfectly fine. No circular reference here. But here's what's in **B1**:

$$=A1 * 2$$

As you can see, **A1** has a formula that refers to **B1**, which in turn has a formula referring back to **A1**. Although the reference is more roundabout, it is circular nonetheless. Once again, the program replies, "Cannot resolve circular references," but goes ahead and inserts a zero in the cell.

Circular references are not necessarily wrong. For some kinds of problems, it can be productive to keep the computer running in circles, so to speak. The trick is to limit the number of circular calculation passes, or *iterations*. Two criteria can control when the circular calculations will stop: *maximum number of iterations* (say, 100 passes) and *maximum change*, whichever occurs first. The maximum change value becomes important if repeating the calculation causes the result to come closer and closer to some value that it never quite reaches. For example, calculation might stop when the next pass produces a difference smaller than 0.001.

Circular references are just one example of flawed spreadsheet logic resulting from one formula using the results of another. The inter-relationship of formulas in a sheet can get very complicated, and that's why makers of spreadsheets must not only be careful designing their sheets but also must devise tests to make sure their formulas work as intended.

Eight Great Steps to Spreadsheet Reliability

Here are some basic steps you can follow to ensure that your spreadsheet (or one designed by a suspicious third party) produces reliable results:

1. Pick the Right Function for the Job

Beware of differences between functions that have similar purposes: INTRATE for securities, YIELD for bonds; ROUND for rounding, TRUNC for truncation; AVERAGE for arithmetic averages, but MEDIAN for the mid-point value.

2. Watch for the Kinds of Errors the Computer Can't Catch

Avoid division by zero, or by any reference that might result in zero.

Beware of any rounded or truncated value that is used in other calculations, particularly multiplication. Inaccuracies will get worse each time an approximated value is multiplied.

Avoid circular references unless you need to solve a problem that requires many recalculation passes (iterations).

Be careful when doing calculations with dates, such as subtracting dates to find out how long an invoice has been unpaid. Computer systems and accounting departments vary in the ways they handle dates: for example, does the date 09/06/99 mean September 6 (U.S. convention) or June 9 (European convention)? Do *serial date values*—the format in which your computer actually stores and works with dates—begin at the year 1900 (IBM PC convention) or 1904 (Apple convention)? Is the accounting year based on the normal 365-day calendar or on a special 360-day year with all 12 months having just 30 days each? Does the accounting year start in January or is it a fiscal year that starts in, say, July?

3. Make Sure to Give a Function Everything It Requires

If you don't enter the arguments as required, the program might be smart enough to reject it. For example, it may respond, "Too few arguments," or "Too many arguments." Or it might put a coded error message in the cell instead of a numeric result. But it has no way of knowing if the cell addresses you gave it actually hold the numbers you want to use.

Be sure to supply the correct *keywords* (special-purpose literal text strings) for the functions that require them, such as **INFO**. Enclose keywords in quotation marks ("")and spell them correctly. For example, the formula

$$=INFO(\text{"MEMAVAIL"})$$

will tell you how much free memory is in the computer.

Enclose text values used as arguments in quotation marks:

$$\text{“JANUARY”}$$

If you omit the quotation marks, the program will think that the item is a range name and will try to find its value.

In arguments, you can use parentheses to control the order in which calculations are done:

$$(2*4+3)=11$$

$$(2*(4+3))=14$$

Parentheses must be used in matching pairs. If you get lost among such *nested* pairs of parentheses, count them. If you have an odd number, you know you have not used the parentheses correctly. But if you have an even number, you can't assume that the rest of formula is correct. Always make sure to check sets of parentheses carefully.

TRUE and **FALSE** are logical values (equivalent to 1 and 0), not text. Do not enclose them in quotation marks when you use them in formulas or when you type them as values in cells.

4. **Verify All References to Data**

Make sure that arguments refer to cells or ranges that hold the values needed by a formula. This is probably the most common source of errors from formulas—especially from formulas that *appear* to work because they produce values instead of error messages.

Use range names, such as

$$\text{SALES} \quad \text{EXPENSES} \quad \text{PROFIT}$$

for clarity, but don't assume that the name refers to the correct range until you recheck it. Beware of copying or moving cells or ranges within a sheet—it can change the references in any formulas it contains. (The program might make adjustments for this automatically, but the results won't necessarily be correct.)

5. Double-Check Your Math

When you're done building a spreadsheet, make a copy of it for testing purposes. (Testing might produce more errors in the sheet, so always test a *copy*.) Enter sample data and note the results. First, use stuff that produces answers you already know or can calculate manually. Then, use data at extremes (high and low anticipated values), and even try data that's out of bounds. Use data of incorrect types, such as a date entered as text where a date value is expected, as well as of correct types.

Watch for error messages and incorrect results. Revise your formulas until the sheet works as expected.

6. Ask: Do These Results *Look* Reasonable?

Generating graphs from ranges in a sheet can be a quick visual check on the probable accuracy of results. (The accounting term for an estimate of probable accuracy is *reasonableness*.) Besides, charts are prettier, and learning how to make them makes good reading.

7. If You Make Changes to the Sheet, Recheck and Retest It

Remember that a spreadsheet program will usually change the cell references in formulas if you copy or move ranges, or if you delete or add columns or rows. Inspect and recheck the formulas after you've rearranged the sheet to be sure that they are referring to the correct sources of data.

8. If You Think Your Computer Goofed, Think Again!

Spreadsheet software can catch the kinds of blatant errors you might make by entering a formula incorrectly. But if your formulas are correctly composed (have correct *syntax*), the computer will slavishly do all your math without complaining. No computer program, however sophisticated, is smart enough to detect errors in your understanding of underlying business problems. Your spreadsheet might appear to work, but only you can judge whether it works right.

Charting Your Way to a Brighter Future

The old proverb that a picture is worth a thousand words goes for numbers, too. Don't publish data in a table or in a spreadsheet that can be shown more clearly and persuasively in a chart. Nowadays, most spreadsheet software has built-in graphing capability—so what's your excuse?

The most important consideration in making a chart from a spreadsheet has nothing to do with computer software: You must exercise some careful judgment in selecting which data will be charted. No graphics software yet invented can make sure that the input data actually describes a meaningful trend or a key relationship. That's why programs that are advertised to generate graphs with a single command or click of a button aren't necessarily a help. You might be able to make a chart quickly (and there's no reason why you shouldn't!), but your understanding of the data, not the ease of making pretty pictures with it, will determine whether the results will make sense to anyone.

A rule of thumb is to be downright picky about the data you use for charting. In any sheet, there is almost always too much information to make a single, understandable chart. Out of the whole sheet, you usually need to find just *one or two rows* of data that tell a story.

Words for Your Charts

Or, Is There Any Truth in Labeling?

INADVERTENTLY **leaving off helpful labels** on a chart is the quickest way to perjure yourself. In Chapter 1, I invoke the memory of my high-school chemistry teacher, Mr. McNeill, who punished me mercilessly (on paper) for omitting labels from numeric answers. In this chapter, I'll show you how data labels can clarify or confuse the interpretation of charts, and I'll expand the discussion considerably to include guidelines for all kinds of explanatory text in presentations—including the text of your speech or report.

Another of my teachers, French philosopher Roland Barthes, said that, syntactically, a book is a long sentence spoken by its author. It might be productive to think of your report as a long caption for your charts. If you say something that doesn't have a picture to go with it, come up with a new picture—or get rid of the material.

Did I Give You Data or Information?

Labels can mean the difference between having raw data and having useful information. To recap Chapter 1, the categories of things on your charts can be indicated with labels, which are as necessary to reliable data as the numbers(or *values)*. As a rule, wherever you see a number, you should find a label nearby.

19 APPLES!

Often a label must have two parts: a unit of measure, and a description of the thing being measured.

19 TRUCKLOADS OF APPLES

Labels on data can sometimes be hidden or implied. Certain conventions for displaying results are shortcuts to labeling data. One type of shortcut involves using a specific style of punctuation to subdivide a set of numbers. To pick an example I've used previously, the following set of numbers has no obvious meaning:

5105238233

But if I set them off by using a particular style of punctuation, you will have some valuable information:

(510) 523-8233

Aided by the helpful parentheses and hyphen, you might recognize this set of digits as a North American telephone number (which, not by mere chance, belongs to the esteemed publishers of this book). In computer jargon, the punctuation marks are *delimiters*, and they mark the divisions between information *fields*. (For an explanation of fields in data records, see Chapter 9.) The hidden labels are the implied names of these fields, which you know by the convention for composing phone numbers:

Area Code	Exchange	Number
510	523	8233

It is a liar's trick to mislabel even the data that has hidden labels by using delimiting conventions that are either unfamiliar to the audience, or incorrect in the context of the presentation.

Here's an example of another set of delimiting conventions, a date written as a set of numbers:

090699

In the United States, an audience might expect to see the date delimited as follows:

09-06-99

or

09/06/99

The implied fields are:

Month	Day	Year
9	6	1999

For an audience in the United States, then, the date will be interpreted as:

SEPTEMBER 6, 1999

But if I use another style of delimiter, you may get a different clue about the interpretation of the date:

09.06.99

The periods are not typically used in the United States for showing dates, but this is a common convention in Europe. And there, the implied labels are different:

Day	Month	Year
9	6	1999

In Europe, the correct interpretation of the same set of digits would be:

JUNE 9, 1999

The example of different date conventions might seem trivial, but it's symptomatic of more sophisticated liar's tricks. For example, when the presenter speaks of results for "the current year," is the reference to the calendar or to the fiscal year? Any confusion about which accounting period is being discussed could have major financial implications, in which case a misunderstanding about date format might not be trivial at all. Any company that does business internationally must establish consistent standards for such simple rules as formats for monetary amounts and dates which appear in reports and presentations.

The Golden Rules of Data Labeling

There are generally two types of labels on charts: those on data points and those on axes. Here are some guidelines for each type.

Labeling Data Points

A basic tool of chart designers is to label the values of data points on a line plot or set of bars:

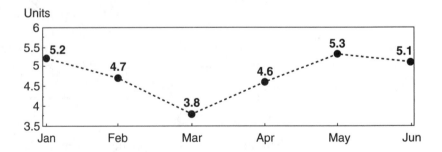

You needn't include data labels on all your charts. In fact, too many labels can clutter the design and actually distract from the interpretation of the graph. But there are circumstances when labeling should be mandatory. In general, you should use data labels to:

- Identify a specific data point you want to highlight in your presentation, such as an exceptionally high or low value, or an average:

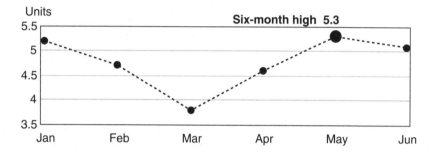

- Make sure that the audience interprets the correct value on plots which are difficult to read, such as estimating the height of 3D bars in relation to the chart scale:

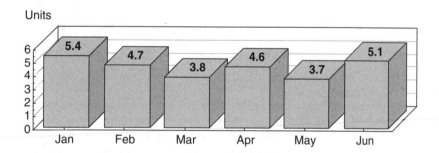

- Show the actual value for a small bar segment or pie slice that had to be made bigger just to be visible:

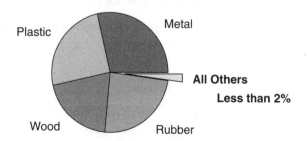

- In an *xy* scatter chart, identify actual values from which trend lines were calculated (such as by extrapolation, interpolation, or regression):

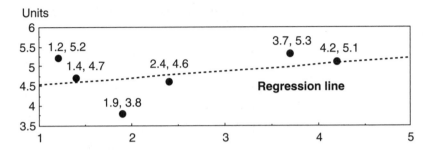

- If the design does not permit locating the label close to the data point, use a *callout line* to direct the eye:

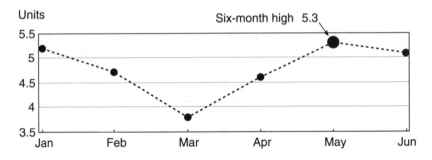

As I emphasize in Chapter 2, you should *not* show labels of actual values on pie charts or in any chart format that shows proportional relationships. If the purpose of the chart is to show proportions, use percentage labels instead. If you must show actual values, do it on a separate chart—perhaps formatted as a vertical bar graph.

Options for Labeling Chart Axes

There are two elements of labeling the axes of *xy* charts: axis titles and scale division labels.

Axis Titles

An axis title simply identifies what type of measurement the axis represents. Here are the rules:

- Include both the type of measurement and the unit of measure in an axis title:

SALES ($)

- You can omit an axis title only if it would be obvious from the scale labels:

MON TUE WED THU FRI

- If a scaling factor, or multiplier, is needed to interpret the scale labels properly, it must be included either in the axis title or in a footnote to the chart.

SALES ($ MILLIONS)

*SALES DOLLARS IN MILLIONS

- If an *xy* chart has dual-*y* axes, you must show an axis title for each. Ideally, the axis titles and legend entries should be color-coded to their respective data series. (See Figure 10.1.)

Figure 10.1 *Charts with dual-y axes should have y-axis titles that are color-coded to legend entries and plots.*

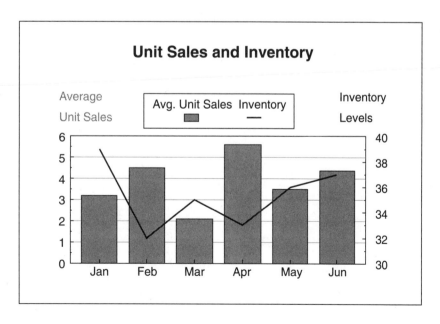

Scale Labels

The honest trick with scale labels is to make them as large and readable as possible. (You'll find some guidelines on choosing type sizes later in this chapter.) Besides increasing the type size, there are two other techniques you can use to enhance readability of scale labels:

- Use a *skip factor* to omit labels and regular intervals from a sequence. For example, if the skip factor were 5, every fifth label would be printed on the scale:

- If after using a skip factor the labels still seem cluttered, adjust their orientation. The normal orientation of labels is horizontal. Some of the alternatives are: vertical up, vertical down, slant up, slant down, and stagger. (Vertical up might still be difficult for some people to read.)

HORIZONTAL

P
U
L
A
C
I
T
R
E
V

V
E
R
T
I
C
A
L

D
O
W
N

SLANT UP

SLANT DOWN

STAGGER   STAGGER

STAGGER

Composing Legends and Data Tables

A chart *legend* correlates the color or pattern of a plot to the name of its data series:

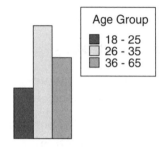

Another term for *legend* is *key*. If there aren't many data series in a chart, an alternative to providing a key is simply to label the data plots:

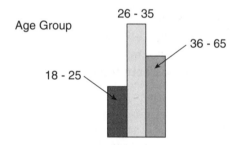

 It's a liar's trick to use color codes that are too similar, making it difficult to correlate the names of data series with their plots.

An alternative to a plain legend is a data table, which shows the numeric values for each data series, usually beneath the chart area. It is customary to show a legend or a data table, but not both. The data table usually incorporates the legend:

Age Grp.	Movies	Television	Newspaper	Other
18 - 25	35	40	23	2
26 - 35	30	42	24	4
36 - 65	26	39	30	5

On the TV quiz show *Wheel of Fortune,* contestants have to buy each vowel to solve word puzzles. Chart liars must think the same way: They use labels so sparingly that you'd think they're being charged a high price for each letter!

Some Notes on Footnotes

In a chart, footnotes can be used to insert citations of sources of data and other explanatory material, as shown in Figure 10.2.

If there is a report narrative accompanying a printed chart, the form of a footnote usually requires a superscripted number after the sentence in the text to which the note refers.[1] In the most common format, the numbered footnote appears at the bottom of the same page.

As far as I know, there are no formal rules of style for footnotes in visual presentations. However, citation of information sources can be a serious matter. In general, you can be guided by the rules for printed publications, where the form of the citation depends on the type of

---

1. Lengthy notes also can be continued on the following pages of a report. However, breaking footnotes between pages is not only troublesome for text composition, but it can also be confusing to the reader.

Figure 10.2 *Use a footnote to disclose the source of information for a chart or the grantor of permission for reproduction.*

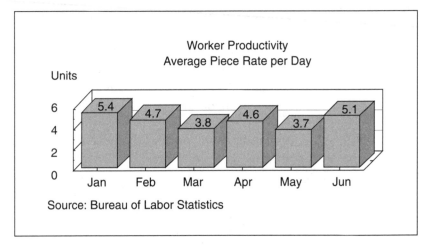

source—whether from a textbook or from a scholarly journal, for example. For the form of footnote references and citation of sources, see Chapter 15 of *The Chicago Manual of Style*, Fourteenth Edition, (Chicago: University of Chicago Press, 1993).

It's Not Just Polite, It's the Law

An important use of footnotes in both reports and presentations is to give credit to the organizations or individuals who hold intellectual property rights to the material. Again, the rules are more formal in print, but you should be aware of some of the legalities involved.

The liar's trick with footnotes is pretty obvious: They don't use them because they either are not willing to disclose the source of their material—or they have no authoritative source at all!

Requirements for obtaining permission to reproduce quoted material and drawings are covered by the 1976 revision of the U.S. Copyright Act. Opinions differ on the doctrine of *fair use*, or reproduction of small excerpts from a longer work without explicit permission. If you get formal permission, you should set the *credit line*, or the citation of the source, exactly as specified by the permission grantor. If you use a short quotation under fair use, the source should still be cited.

Credit lines for photographs and illustrations are generally set in the figure captions. Permissions to quote may be included on the copyright page of a long report, which in a book should be on the back of the title page. Permissions might also be included in a separate section of acknowledgments. In more formal documents, such as scholarly articles and books, citations are included in footnotes at the bottom of the page or numbered by chapter in an appendix titled "Notes" or "References." (Be aware that a bibliography is not a list of specific citations, but a separate listing of research sources.)

Related to the subject of permissions is the question of plagiarism. Don't assume that you can get around the need to obtain permission simply by paraphrasing, or rewording, a quotation or by redrawing a chart or illustration. The revision to the copyright act uses the standard of *substantial similarity* to define plagiarism—and leaves it to the courts to decide what that means in each case.

Some material is so old that it is out of copyright, or in the *public domain*. This material might not require permission for quotation. But be careful! For example, no one holds a copyright on the words of Shakespeare. However, the *typesetting* of his words in a new book edition might be protected by copyright law—so you can't just go reproducing the pages. Also protected would be a new translation into another language, scholarly comments or references included in the text, illustrations, and so on. Likewise, no one has the rights to Leonardo da Vinci's drawings. However, a museum typically would restrict public access to the originals, and a person who was granted permission to photograph those drawings might claim proprietary interest in the right to reproduce the photographs.

Clip art is predrawn art sold specifically to be incorporated into your presentations. However, publication rights vary among vendors. If you plan to offer a report for sale to the public, consult the language of the license agreement that came with the clip art library.

Don't look here for an interpretation of the law. I'm not a lawyer, but I reserve the right to play one on TV.

Readability Is Fundamental!

It's fine to talk about making chart labels readable, but you can't do that without some specifics on type composition. Here is an overview.

What Type of Type Is a Font?

To a professional typesetter, a *font* is a particular design of lettering in a specific size. The design is called a *typeface* (or just *face*).

Type size is usually measured in *points*. There are 72 points to an inch. (The exact specification used in typesetting is 72.254 points to the inch, but who cares?)

A standard typeface on typewriters is Courier. Although desktop computers can produce a wide variety of more attractive type styles, many businesspeople still expect to see formal reports printed in

Courier. Courier Elite on a typewriter is the same as 10-point type, and Courier Pica is 12 points:

```
Courier Elite

Courier Pica
```

All Upstanding Fonts Have Character

A font generally includes all of the letters of the alphabet in capitals (A–Z) and in lowercase (a–z), as well as the numeric digits 0–9, punctuation marks, special characters such as the dollar sign ($), and perhaps some characters used in other languages (usually referred to as *international characters*). All of the characters in a font constitute its *character set*.

All of the characters in the Arial font are shown in Figure 10.3. This Microsoft Windows font is similar to Helvetica, which is probably the most widely used set of letters in the world.

Fonts differ in the number and variety of characters that they contain. For example, some special-purpose fonts have only capital letters. Most English-language fonts include the British pound sign (£), but many of them do not have the copyright symbol (©).

The Short Story of the Points and Picas

Standard single-spacing on a typewriter is 6 lines to the inch. This is also the definition of the unit of measurement called the *pica*. In conventional typesetting, the point unit is used with type, and the pica is used for measurements on a page, such as leading and the widths of lines and columns.

However, on most personal computers, *all* numeric settings—including type size, letterspacing (space between letters in a word) and leading (vertical space between lines)—are in points. Vertical spacing is usually controlled as a multiple of the height of a line, which depends on the point size of the type: The bigger the type, the more space between the lines. So, single-spacing will produce 6 lines per inch *only if the type size is 12 points* (6 lines/inch × 12 pt = 72 points/inch). It should come as no surprise, then, that 12-point typewriter type is called Pica: Each character is one pica tall.

Figure 10.3 *The character set of the font Arial*

	!	"	#	$	%	&	'	(	)	*	+	,	-	.	/	0	1	2	3	4	5	6	7	8	9	:	;	<	=	>	?
@	A	B	C	D	E	F	G	H	I	J	K	L	M	N	O	P	Q	R	S	T	U	V	W	X	Y	Z	[	\	]	^	_
`	a	b	c	d	e	f	g	h	i	j	k	l	m	n	o	p	q	r	s	t	u	v	w	x	y	z	{	|	}	~	□
□	□	,	ƒ	„	…	†	‡	^	‰	Š	‹	Œ	□	□	□	□	'	'	"	"	•	—	—	~	™	š	›	œ	□	□	Ÿ
¡	¢	£	¤	¥	¦	§	¨	©	ª	«	¬	-	®	¯	°	±	²	³	´	µ	¶	·	¸	¹	º	»	¼	½	¾	¿
À	Á	Â	Ã	Ä	Å	Æ	Ç	È	É	Ê	Ë	Ì	Í	Î	Ï	Ð	Ñ	Ò	Ó	Ô	Õ	Ö	×	Ø	Ù	Ú	Û	Ü	Ý	Þ	ß
à	á	â	ã	ä	å	æ	ç	è	é	ê	ë	ì	í	î	ï	ð	ñ	ò	ó	ô	õ	ö	÷	ø	ù	ú	û	ü	ý	þ	ÿ

When Is a Font Just a Font?

Except to graphic arts professionals, the term *font* usually means just *typeface*, without reference to size. That's because most of today's computer fonts are *scalable*, or can be adjusted continuously over a wide range of point sizes, typically from 1 (very small) to 999 (very big).

In software manuals and in books like this one, you will sometimes see the term *font* used rather loosely to refer to the whole set of appearance options, or *attributes*, that are available for a particular typeface.

Attributes Are the Spice of Type

Attributes are ways of varying the appearance of a typeface. Font attributes can be applied to individual characters, or letters—or to groups of letters, such as words or even entire paragraphs or documents. Examples of font attributes include bold, italic, underline, double underline, strikethrough, color, hidden, small caps, all caps, spacing, superscript, subscript, and drop shadow. Most of these attributes are shown in Figure 10.4.

Knowing about font attributes is particularly useful when you are composing charts for a printed report that will be reproduced in black-and-white. Attributes such as bold and underline can be used instead of color to highlight labels and emphasize text.

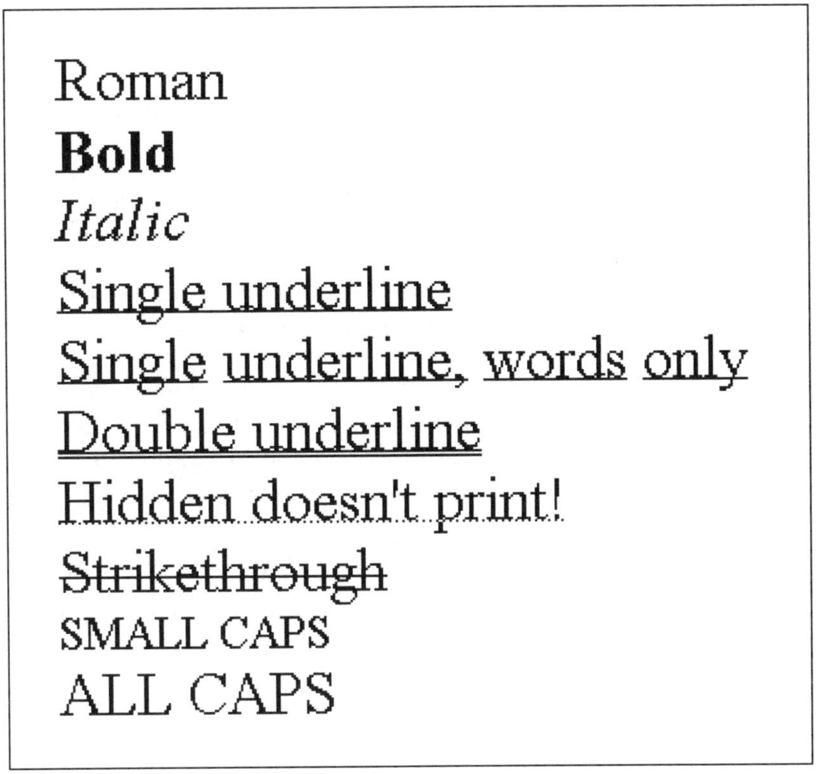

Figure 10.4 *Font attributes provide many ways to alter the appearance—and the impression given by—a single typeface. The font used in this example is Times New Roman.*

Roman

Bold

Italic

<u>Single underline</u>

<u>Single</u> <u>underline</u>, <u>words</u> <u>only</u>

<u>Double underline</u>

Hidden doesn't print!

~~Strikethrough~~

SMALL CAPS

ALL CAPS

Shopping for Just the Right Text Font

There are no strict rules about choosing fonts for your presentation, but here are some tips I hope you find valuable:

Use a proportionally spaced font.

For attractive letterspacing and good readability, pick a *proportional* font—one that varies the space between the letters—for more attractive composition. There's just no good reason to use a *monospace* font—one with the same amount of space for each letter—for presentation-quality text. Monospace fonts can simplify alignment of numeric digits in the columns of a table. However, for precisely this reason, the numeric digits in proportional fonts are usually monospace, not proportional.

Proportional (Times New Roman)

```
Monospace (Courier)
```

Use sans serif fonts for chart labels, serif fonts for titles and explanatory text.

Plain, or *sans serif*, fonts like Arial and Helvetica are standard for chart text, such as axis and data labels.

The serifs, or curlicues and doodads, at the ends of letters can enhance the legibility of text, as well as add a distinctive impression. You might use a serif font, such as Times or Times New Roman, for chart titles and footnotes.

Don't intermix fonts within a chart.

You might use a serif font for the title, subheadings, and footnotes, and just one sans serif font for all the chart labels. Intermixing fonts within the chart itself will probably be a needless distraction for your audience.

Use upper- and lowercase letters.

While readability refers to the ease of recognizing words and sentences, legibility refers the ease of recognizing individual characters, or *letterforms*. Legibility is increased if you use upper- and lowercase letters. In particular, the *ascenders* and *descenders* of lowercase letters are an aid to quick recognition.

Ascenders	Descenders
b d f h k l t	**g j p q y**

Liar's Don't Care About Readability on the Screen

If your visual presentation will be viewed on a monitor or projected on a wall screen, remember that the typical viewing distance is *four times the height of the screen*. Therefore, don't use text smaller than can be read comfortably at this distance.

Monospace Fonts: An Endangered Species

Are monospace fonts threatened with extinction? The only reason for the persistence of Courier and its clones in business correspondence is that people have entrenched expectations that go back at least as far as the 1960s and the first IBM Selectric typewriters. As far as I know, there are only two reasons to use monospace fonts—ever:

1. When you are building a table, monospacing makes it easy to align the entries vertically. (See Chapter 9 on setting a nice table!)

2. If you are writing a report or magazine article and must be concerned with the word count, monospacing can help. The number of characters per page will be more consistent than with proportional type. However, since most word processing programs can count the words for you, what's the point? (Some programs give only a count of characters in a document. To get the word count, divide the character count by 8, unless you're using every long word in your vocabulary to write a scholarly dissertation.)

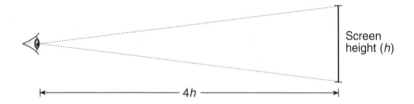

According to the Society of Motion Picture and Television Engineers, the minimum height of letters used in on-screen text should be at least 2.5 percent of the height of the screen. The height of a 14-inch computer or video screen is 8.25 inches, or 594 points (72 points per inch). So, the characters should be at least 2.5 percent of 594, or 14.85, points tall. However, remember that this measurement is *not* the type point size. The point size of a font is measured from the bottom of a lowercase descender to the top of an ascender, as shown in Figure 10.5. The specification here is for the minimum height of any letter, such as a lowercase *a*. The actual point size of a font for which the lowercase *a* is about 14 points tall would be about 20–24 points, depending on the design of the font.

Figure 10.5 *Don't confuse the minimum height of characters with the point size of the font.*

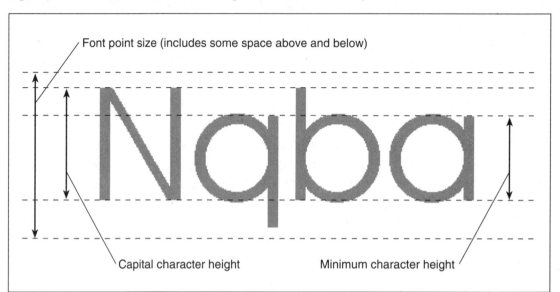

To further promote readability on the screen, avoid large areas of white as backgrounds. In general, use dark backgrounds with light-colored text for projected media, such as slides. Adjust these colors for the typical lighting conditions in the room. The brighter room light, the lighter the background color must be. In a highly lit room, you might need to use a light-colored background with dark text.

See Chapter 11 for more about use of color in presentations.

Just Your Style (Short Version)

Another mark of an accomplished liar is a studied sloppiness about punctuation and grammar, especially if the result is an ambiguous meaning or one that is downright wrong!

Here are some of the generally accepted rules of style that you can use for checking your presentation—both its charts and its report narrative. Some business organizations—and most publications—have their own manuals of style, which cover many of these topics. Even if your company has a style manual, you will have to adhere to the editorial policies of the magazine or journal in published articles.

Punctuation

These aren't all the rules—just the ones that get missed most often:

Apostrophe

Generally, use *'s* to indicate the possessive form of a noun or indefinite pronoun, or add just the *'* to a plural form that ends in *s*. However, if the word ends in an *s* or *z* sound but is singular, add *'s* if this last syllable will be pronounced, but just the *'* if the possessive part isn't pronounced:

> ## The box's shape
> ## For goodness' sake
> ## Mills' report
> ## Mr. Ness's presentation

Colon

You can use a **:** instead of *and* to join a compound sentence. However, the second clause should amplify or clarify the first. Style books differ on whether to capitalize the first word of the second clause:

> ## This is certain: we will have to penetrate new markets next year.

or

> ## This is certain: We will have to penetrate new markets next year.

There is more reason to capitalize the second clause if it is lengthy.

Comma

The *serial comma* is often a point of contention. Is this sentence punctuated correctly?

We sell hardware, software, and service.

Or do you hold that the *and* takes the place of that last comma in a series:

We sell hardware, software and service.

I'd say experts on style are split about 50-50 on this one. Personally, I'm in favor of the serial comma, especially when the *and*'s start to proliferate:

We sell hardware, software, software support and hardware maintenance, and custom programming.

Another use of commas is to set off dates and geographic locations when they appear in the middle of a sentence:

We are located in Boulder, Colorado, but our home office is in Sioux City, Iowa.

On June 6, 1999, the note will become due.

However, if the reference includes *only the month and year*, the commas are omitted:

The note will come due in June 1999.

Dash

If a long dash (called an *em* dash) is used to separate two clauses, the comma that would otherwise go there is omitted:

Our technical advisors cannot afford to make mistakes—our customers will be watching us closely.

Ellipsis

This one is tricky. Use three dots (…) to indicate words that have been omitted from a quotation if the omission is *within* a sentence. If the omission is at the end, you need *four* dots—three for the ellipsis and one for the period! (Use the font's ellipsis character rather than a string of periods.)

Hyphen

Use hyphens in compound words (*self-examination*) or to divide broken words at the ends of lines. (Remember that the hyphen is being dropped increasingly from compound words, even if it means joining two vowels: *reenter.*) Use an *en dash* (slightly longer) instead of a hyphen to indicate a range (0–60 mph). Also, don't use a hyphen for a minus sign (–).

Double-hyphen

If the word you break at the end of a line in a report narrative was hyphenated at that point already, you should use a double hyphen (⸗), but you won't find it in most fonts. (This rule is seldom followed anymore.)

Parentheses

If an entire sentence is enclosed within parentheses, capitalize the first word and put the period inside the closing parenthesis:

> (Information in this report is based on unaudited results.)

But, if the parenthetical sentence appears inside another sentence, omit the initial cap and the period, although a question mark or exclamation point is permitted:

> This fact is obvious (do we ever question it?), but it is a dangerous assumption.

If the parenthetical expression is a fragment and it comes at the end of the sentence, put the period outside of the closing parenthesis:

> This much is obvious (or maybe not).

Period

Style experts differ in their requirements for periods after some abbreviations. In general, use periods if the abbreviation is not an accepted acronym, but don't expect that rule to apply in all cases.

> made in the USA
> Meet me at 10 A.M. EST.
> the Hon. Thurgood Marshall

Question Mark

Here are two different ways of posing a series of questions:

> I want to know which product will you specify? what size will you want? do you prefer white or green?

I want to know (1) which product will you specify, (2) what size will you want, (3) do you prefer white or green?

Quotation Marks

Here's another item that you'll often see done incorrectly. Periods and commas go *inside* the closing quotation mark—even if not part of the quoted material. Colons and semicolons go outside. Dashes, question marks, and exclamation points go inside if part of the quotation, outside if part of the sentence as a whole.

If a quotation contains another quotation, enclose the embedded quotation in single quotation marks ('). If the embedded quotation comes at the end of the sentence, put the period first, then the single quotation mark, then the double:

Her exact response was, "I said, 'Don't even think about bringing that up at the meeting.'"

Semicolon

You can use a semicolon in place of *and* to join a compound sentence, omitting the comma. But use the semicolon in place of the comma and keep the *and* if either clause contains commas:

Great Britain, France, and the United States were allies in both world wars; and, in the midst of intense economic competition, those ties still hold.

The formal style books also say to use a semicolon to separate compound items in a series:

> ## The color schemes are red, white, and blue; orange, green, and brown; and blue, blue-green, and yellow.

However, some writers, especially popular journalists and novelists, are doing away with using semicolons in place of commas (omitting some of the commas, as well), with the idea perhaps that the flow of the sentence can be improved:

> ## Great Britain, France, and the United States were allies in both world wars, and in the midst of intense economic competition those ties still hold.

Use of Italics

Use italics for emphasis; to highlight letters, words, and phrases referred to as such; in foreign words and phrases; and for titles of longer works such as books, manuals, and names of periodicals.

Use double quotation marks rather than italics for titles of articles within periodicals, as well as for chapters or sections within books.

Experts differ on which words are considered foreign and should therefore be italicized. A good working rule is that if you can find it in the A–Z part of the English dictionary, it need not be italicized. Examples are the Latin words versus and de facto, which are in everyday speech. But you might italicize a term if you think your readers won't know it:

> ## She spoke of the *zeitgeist* of people at a dynamic company during a period of rapid change.

A common area of confusion is whether to put special words or phrases in italics or to enclose them in quotation marks. In the cases just cited, italics are correct. Use quotation marks (sparingly) to enclose words that are used in unusual ways. Avoid using quotation marks to give special emphasis to words that are used conventionally.

Capitalization

There are lots of capitalization pitfalls. Here are just a few of them:

Quoted Material

Capitalize the first word in a quotation *unless* the quotation completes the sentence:

> The defendant said, "I made the whole thing up."

> The defendant insisted that he "made the whole thing up."

Bulleted Items

The first word of bulleted items should be capitalized, whether or not the item is a sentence.

Headlines and Titles

This is an important one for publications and a point of some confusion.

Capitalize each word in a headline or title except for internal conjunctions (*and,* unless it's the first word), prepositions (*in, with*), and articles (*a, the*).

But—depending on the style expert—longer prepositions are also capitalized:

Choosing Between These Two Alternatives

One rule is, if the word has more than five letters, capitalize it.

Titles of Persons

Government and corporate titles are capitalized only if used with the proper name of the person:

President T. J. Mackenzie spoke.

The president spoke.

Some government officials can be addressed by their title in place of their names:

Mr. President, take a look at the polls!

Thank you, Mr. Speaker.

Numbers and Fractions

In a narrative and in chart notes, spell out numbers from *one* to *nine* unless they are used with units of measure:

We have two, perhaps three, choices.

The machined part will require a taper of 3 cm.

Spell out the commonly used fractions *one-quarter* and *one-half,* unless the fraction is preceded by a whole-number digit or followed by a unit of measure. Omit the endings *st, rd, nd, th,* or *ths* after numeric digits or fractions. Do not use *of a* or *of an* between a fraction and its unit of measure ($\frac{1}{2}$ inch).

As I discuss in Chapter 1, use the typeset fractions included in a font rather than constructing them from two digits separated by a slash ($\frac{1}{2}$ rather than 1/2). And don't put a space between a whole number and the fraction that follows it ($5\frac{1}{4}$).

Alternate Spellings

Some people have specific preferences about alternate spellings, hoping perhaps to revolutionize the language single-handedly. For example, both of these spellings are correct, but a company style guide or publication should enforce the use of only one of them:

judgment

judgement

Where there are two correct spellings of a word, the preferred spelling is listed first in the dictionary. (But, which dictionary? And preferred by whom?)

Less obvious choices involve technical terms and jargon that are just starting to be accepted in general usage. Here are two versions of a familiar term that are now considered equally correct:

file name

filename

Until recently, you had other options for *workstation*—work-station and work station. Both are now considered incorrect.

A manual of style used by a corporation or publication will usually include a list of preferred spellings—especially for the technical terms and jargon of the trade or profession—and will also cite a specific edition of a dictionary to be consulted for anything else. The most commonly cited authority in the United States is *Merriam-Webster's Collegiate Dictionary* (Tenth edition).

If you have to go to an unabridged dictionary to find a word, many of your readers might not know it. Consider using a more common synonym or even several simpler words that describe the concept.

In general, abbreviations should be avoided within sentences, but exceptions might be permitted for units (MHz for megahertz), geographical names (Calif.), forms of address (Ms. with or Ms without the period), and so on. Style manuals should list the preferred forms of abbreviations and acronyms that are commonly used in the organization or publication.

Person and Voice

This category has less to do with charting but is nevertheless an important section of any corporate policy on report writing.

Depending on the formality of the report or presentation, a style manual might have rules about direct references to the author or to the reader. Some publications permit *I* and *you*, although many discourage *we* as condescending or confusing (who are *we*, anyway?). The "editorial *we*," with which the author attempts to hide in an anonymous group (with the corporate staff? with fellow wizards?), has all but disappeared from contemporary usage.

LIAR'S TRICK

Liars are prone to the inclusive *we* as a way of including themselves with known experts or reliable authorities. Corporations are beginning to frown on the use of *we* in formal publications. Lest they appear condescending, presumptuous, or even adversarial, internal corporate publications must be particularly careful about this. It is often a matter of strict policy that *we* cannot be used unless it is clear from the context that the reader is included!

Some more formal publications recommend avoiding personal pronouns except for the third person—*he, she, it,* and *they.* However, this rule can force the author into stuffy formalisms such as

The reader will discover that...

A poor alternative is the cold impersonality—and vagueness—of the passive voice:

It will be discovered that...

Using the passive voice is all the more undesirable because most style manuals encourage writers to use the active voice wherever possible:

You will discover that...

Reading Level

Also related to style is the educational level that is assumed of the audience. Major-market newspapers assume a tenth grade reading level—which is also the standard for most undergraduate college textbooks!

Some grammar-checking software will analyze word-processing files and report the reading level as a score called the *readability index*. But here are some general guidelines, for which you don't need a grammar-checker:

- Avoid words having more than three syllables.

- Keep sentences to less than 20 words.

- Don't include more than two clauses in most sentences.

Want to Be Even More Stylish?

If you want to get serious about this style business, here are some of the most frequently consulted sources:

- Bernstein, Theodore M., *The Careful Writer: A Modern Guide to English Usage* (Macmillan, 1977) ISBN 0-689-10038-8.

- Editorial Staff of the University of Chicago Press, *The Chicago Manual of Style* (Fourteenth edition; University of Chicago Press, 1993) ISBN 0-226-10389-7.

- Fowler, Henry W., *A Dictionary of Modern English Usage* (Second edition; Ohio University Press, 1987) ISBN 0-19-869115-7.

- "Handbook of Style" in *Merriam-Webster's Collegiate Dictionary* (Tenth edition; Merriam-Webster, 1993) ISBN 0-87779-708-0.

- Strunk, William Jr. and E.B. White, *Elements of Style* (Third edition; Macmillan, 1979) ISBN 0-02-418220-6.

- Warren, Thomas L., *Words into Type* (Fourth edition; Prentice-Hall, 1992) ISBN 0-13-966060-7.

Other Quality-Control Steps

In addition to enforcing these standards, the writer of a report or the maker of a chart should review the material for *sense*, as in,

"Will this make sense to most of my audience?"

Although the content of your presentation may be reviewed in advance by your management, you should be responsible for assuring its accuracy. An important step is *fact-checking*—looking up statistics in reference books or calling people who are quoted in an article or report to confirm the accuracy of the quotation.

Business Presentations Aren't Movies!

Oral presentations have a special set of rules. Whether you speak extemporaneously or give a prerecorded pitch, you must explain and interpret a sequence of pictures for the audience. You will be asking for trouble if you throw a chart of last quarter's sales results on the screen and then pause in silent meditation, attending the wisdom of the viewers. You must tell your audience what conclusions *you* draw from the data.

As the presenter, remember that the agenda should be yours. If your objectives aren't clear, or your conclusions aren't specific, you will lose control of the meeting, as well as your influence over the management decision you seek.

A time-honored technique for visualizing the structure of a presentation is to put the outline of your speech on the screen.

Sounds dull, you say?

I don't mean the whole outline, and not all at once. Rather, you can break up the points of a topic outline into text screens, which can give structure to a presentation. Show main topics as titles, and subtopics as lists.

Don't Sit at This Table!

A third type of text screen—the dreary table—is to be avoided. If you are tempted to show data in a table (or worse, if you try to put a whole spreadsheet on the screen), consider converting the numbers to a graph, which will almost always show the relationships better. For tips on creating charts—as well as tables, if you must—see Chapter 9.

Titles Are Topics

A title screen shows the text of a main topic, or title, sometimes with a subtopic, or subtitle, as shown in Figure 10.6. Insert a title screen before each section of charts in your presentation, just as you would introduce the paragraphs of a report with headings and subheadings.

Speechwriters call title screens *signposts*, for the obvious reason that a title points the way to what's ahead. The purpose is to reinforce the structure of your presentation so that the audience is with you at every step.

Figure 10.6 *A title screen tells the audience what to expect, and follows the structure of main topics in your presentation outline.*

How I Spent
the Summer
by Vincent Van Gogh

Impasto Enterprises, Inc.

Put Your Agenda in a List

A set of related subtopics can be presented as a list, with the items bulleted or numbered, as shown in Figure 10.7. A list should recap the subordinate points of one main topic in your presentation outline.

However, showing a naked list (all items on the screen at once) encourages the so-called "read-ahead" problem. The audience can read faster than you can talk. They can see the whole list at a glance and will have a natural tendency not only to read ahead, but also to think ahead—perhaps anticipating your conclusions, perhaps wrongly!

The solution to the read-ahead problem is to subdivide a list into several separate screens as a *build-up sequence*—as shown in Figure 10.8. In choosing colors for the text, you can subdue, or dim, the previously presented items and highlight the current item. The subdued color might be a lighter shade of the background, and the highlight color should contrast sharply with it. So, if the background were dark blue, the subdued color might be light blue, and the highlight color might be white or yellow.

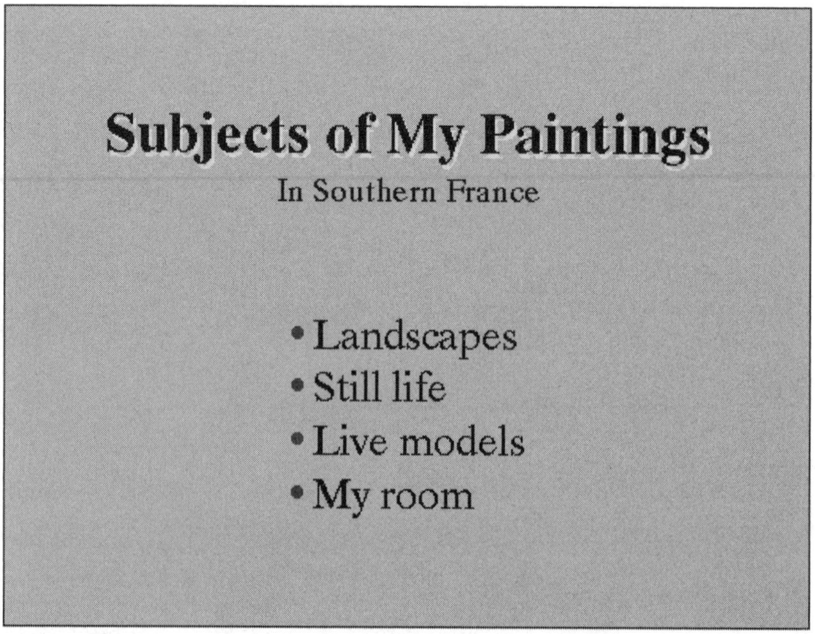

Figure 10.7 *A list recaps the details of one main topic in your speech outline.*

Subjects of My Paintings
In Southern France

- Landscapes
- Still life
- Live models
- My room

Figure 10.8   *A build-up sequence presents a list in stages to avoid the "read-ahead" problem.*

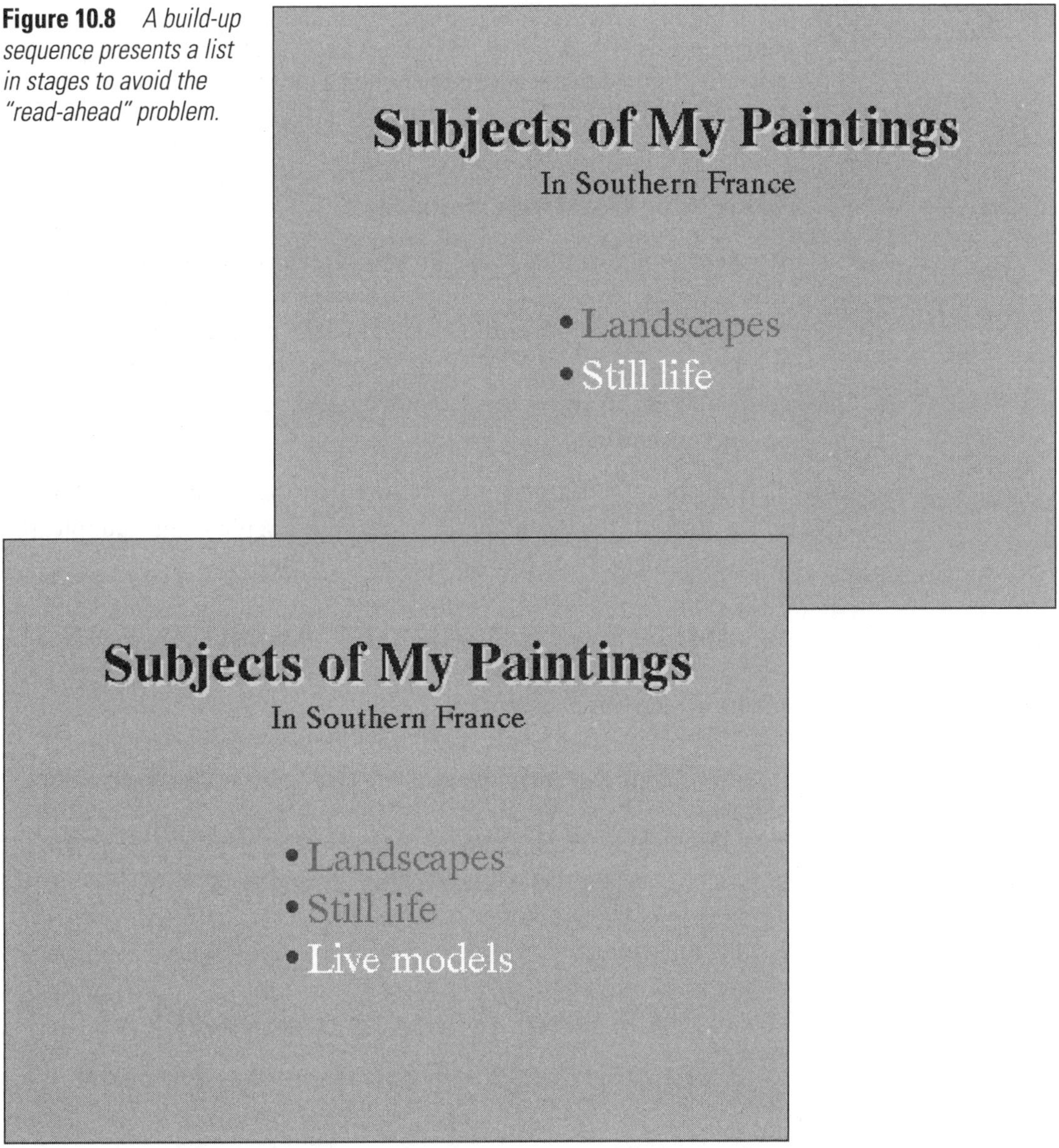

Subjects of My Paintings
In Southern France

- Landscapes
- Still life

Subjects of My Paintings
In Southern France

- Landscapes
- Still life
- Live models

 A liar might not bother to signpost a presentation. After all, if the audience gets lost—so much the better! Or, how about including sign-posts for topics that are never discussed? (I guess I just slept through that part about the money-back guarantee.)

Take Your Cue from Your Text

If you're going to put words on the screen, you'd better say them. And you've got to say them when the audience *sees* them. This synchronization of video with audio is called *visual cueing*. A picture should be cued, or should appear on the screen, at the moment that you begin to talk about it. And if the screen contains text, you should start by saying those words.

Careful speechwriters even make sure that the syntax, or word order, of the spoken words matches the text on the screen. For example, the audience might be distracted needlessly if they see

INCREASE MARKET AWARENESS

on the screen but hear you say,

"Recognition of our product in the marketplace should be enhanced in the coming months."

For the most effective reinforcement of your message, you should say,

"We want to *increase market awareness* of our product in the months ahead."

Find a way to visualize all the important topics of your presentation and put them on the screen. Show key relationships in dramatic charts. But remember—those charts won't be persuasive without your honest and sincere words to back them up!

Color

Its Uses and Abuses

As with so many other lessons in life, things that are pretty can be especially deceiving. Applying color to your charts might be aesthetically pleasing, but looking pretty shouldn't be your only thought. Color is a powerful tool for communication—and by now you should appreciate that any graphic tool holds the potential for abuse by shameless prevaricators. In fact, some accounting professionals are so fearful of the potential misuses of color that they advise sticking to black and white!

Your world needn't be so dull. Go ahead, splash on the color. Just be aware of the attentions you invite when you try to dazzle your admirers with it.

As in personal grooming or interior decorating, you need to develop a sense of style when using color in your charts. As with the other graphic tools I've discussed, learning to use color effectively involves both knowing the expectations of your audience and understanding the circumstances that surround your presentation. You need to know subjective things like people's biases about color, as well as some objective facts about how colors are reproduced in different media— such as paper, film transparencies, and video.

In the realm of color, there's a lot for the ambitious liar to exploit. And, as with subjective notions such as screen direction (left-right, up-down), our prejudices about color are so ingrained that most of us aren't aware of them.

You can also think about color as an aspect of product design—the product being the presentation you're preparing. Initially, your natural inclination will be to please yourself, to make selections that suit your personal taste. But there will almost always be other opinions besides your own that you can't ignore. Your priority should always be meeting the expectations of your audience.

The situation gets more complicated if you are preparing a presentation that will be delivered by someone else—your boss or perhaps a client. Here you might run into the touchy situation where your client's preferences about color run counter to his or her best interests—as far as getting a desired result from the audience is concerned.

A factor that can affect all these sets of opinions is corporate style. Most business organizations have adopted specific sets of corporate colors, possibly also including required typefaces and other graphic rules intended to create a unique, identifiable image. Often these rules are codified in style manuals, which must be followed in preparing any major presentation to management or the public. As a result, if different companies are involved, you might need to create separate sets of colors and presentation styles for each project or client.

Most people don't have to be so exacting about color. They needn't worry, for example, that referring to the *shade* of a color is not the same as describing its *tone*. But if you want to develop a vocabulary for using color in presentations, you must pick your words more carefully.

Color Terminology: The Short Course

The word *color* is too all-encompassing to describe the rich variety of visual impressions that surrounds us. Just as Greenlanders have many different words for *snow*, graphic artists have developed more precise terms for dealing with color. Here are some basic definitions. (The mixing of these elements to produce different colors is diagrammed in Figure 11.1.)

Figure 11.1 *To the discerning graphic artist, what the rest of us call a* color *is a* tone *that has at least six different contributing factors.*

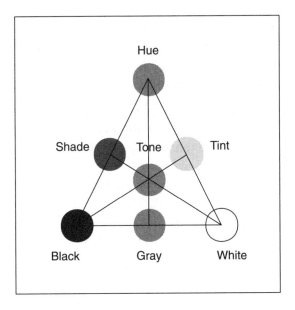

Hue

Describes the primary colors from which a specific color blend, or tone, can be made. The basic hues are red, red-orange, orange, yellow-orange, yellow, yellow-green, green, blue-green, blue, blue-violet, violet, and red-violet.

Tint

Refers to the amount of white that is mixed with a color. Light tints appear washed out. A technical term for tint is *chroma*, or color purity. Some color experts prefer to call it *saturation*.

Shade

Is the amount of black, or darkness, in a color. Similar terms are *value*, *lightness*, and *intensity*.

Tone

Is what in everyday speech you'd call a *color*—the combination of hue, tint (saturation), and shade (value).

If these distinctions don't make much sense to you, walk over to the nearest color TV set. To change the color (tone) on the screen, you can adjust three different controls: Color (hue), Tint (tint), and Brightness (shade).

The technology of a video monitor on a personal computer is not the same as television. The three elements of computer-generated colors are Red, Green, and Blue, or *RGB*. But trying to get the color combination you want by adjusting R, G, and B controls isn't so easy. So, most computer graphics software translates RGB into a set of controls more like the color TV set: Hue, Saturation, and Lightness, or *HSL*. In some programs, these terms are called Hue, Chroma, and Value (HCV), or Hue, Saturation, and Value (HSV).

The Psychology of Color

Colors not only mean different things to different people, but the meanings can also be different depending on the situation those people are in!

At the most subjective, nonverbal level is the question of *mood,* or emotional response. Consider the variations in mood of some basic color hues for different audiences. A survey of American movie audiences reveals one set of moods, as shown in the second column of Table 11.1.

Similar tables are used by movie technicians to adjust the overall tint of scenes to better fit the mood the director intended. Now, you might think that such minor differences don't have anything to do with the bold colors in business presentations. But I'd say that the effect is analogous—in the case of the movies, the color difference is very subtle, and so is the effect on the audience. Make the colors bolder, and you will magnify their effect.

Now look at the third column of Table 11.1. There's a completely different set of preferences for financial managers, based on how they ask professional designers to create presentations for them.

Table 11.1 *Subjective Interpretation of Color by Different Audiences*

Color	Movie Audience	Financial Managers	Health Care Professionals	Control Engineers
		INTERPRETATION		
Blue	Tender	Corporate, reliable	Dead	Cold, water
Cyan	Leisurely	Cool, subdued	Cyanotic, deprived of oxygen	Steam
Green	Playful	Profitable	Infected, bilious	Nominal, safe
Yellow	Happy	Highlighted item, important	Jaundiced	Caution
Red	Exciting	Unprofitable	Healthy	Danger
Magenta	Sad	Wealthy	Cause for concern	Hot, radioactive

The differences don't stop there. Doctors and other health care professionals, shown in the fourth column of the table, don't agree with the financial types.

In the fifth column is another set of reactions from a group of control engineers with responsibility for monitoring a nuclear reactor.

It should be apparent from the table that some biases about color have to do with a person's professional and cultural background. But notions about color aren't that simple: Our interpretation of colors can change with the situation. People make these mental shifts frequently and without conscious effort, usually depending on their understanding of the *purpose* of a visual message.

For example, in some circumstances, it will be natural for movie audiences to react more like control engineers. When the moviegoers exit the theater parking lot in their cars and approach the first traffic signal, their situation will be much closer to that of the control engineers in a power plant: Very quickly, they must interpret a limited set of color codes. In this simplified set of codes, a red light always means *stop*, regardless of how the person might "feel" about red in some other situation.

On the other hand, people with specialized biases, like the control engineers, do not necessarily carry them into other aspects of their lives: A control engineer's reactions to colors in the movie theater are not necessarily any different from those of doctors, accountants, or the public at large.

What does this mean to you as you design a business presentation? Again, you must tailor the presentation to the biases of your audience. Be aware of these biases. To an accountant, green money is good and red ink is bad. Doctors want it just the other way around: oxygen-rich red blood is very good, green necrosis very, very bad! But the situation is also important. The accountants and the doctors will agree on what to do when they see traffic lights on the drive home. And individual tastes and moods will vary even more, for all kinds of subjective reasons—such as the color of the sky on the day you make your pitch!

Liars are skilled at sensing the mood of an audience and then manipulating it to their own ends. A keen sense of how colors affect moods is therefore an essential skill. Yes, perhaps using corporate blue is usually a safe plan. But who knew that using a lot of red in making an investment pitch to a group of doctors would stimulate more excitement than alarm?

What Do You Mean by That Color?

Creating a mood with color involves making an overall impression, perhaps an unconscious one, on the audience. But beyond mood, which can arise even from unintentional color choices, there can be deliberate and obvious uses of color in presentations. That is, color can be used for specific communication purposes.

Color can be used for 1) a kind of coding, 2) rendering of real-life pictures and artistic expression, 3) as a design element of graphic materials, including charts. Business presentations often include all three types of uses.

Color as Coding

The situation of the control engineers in the nuclear plant is an example of color used as coding. Computer displays that monitor conditions within the reactor may use specific colors to signal the status of each of the plant's subsystems.

The power of any coding scheme is its simplicity. When colors are used as codes, you must *limit the choices to the fewest colors necessary*. The greater the number of colors, the more likely your audience will become confused.

For example, think about how confusing it would be to deal with traffic signals that displayed six or seven colors. Instead of GO … CAUTION … STOP, you might be confronted with color codes for:

GO

NO LEFT TURN

NO RIGHT TURN

NO U TURN

CAUTION

WAIT FOR PEDESTRIANS

When colors are used as codes, the precise shades or tints of the color hues probably won't be important—as long as the audience can easily distinguish one code from another.

In developing codes, remember that some people are color blind. The common difficulty is distinguishing opposite colors on the color wheel (more about that later). Red and green might both look like gray or some other neutral color. A solution to this problem is to code the objects two ways—by pattern as well as by color. (This is sometimes done on traffic lights. If you look closely, you might see a bar across the red lens.) So, even if you code two lines in a chart separately by color as red and green, you might also distinguish them as solid and dashed, and areas as solid and patterned:

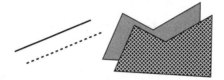

An example of color used as coding is the legend of a chart. The legend provides a key that matches colors, patterns, or both to bars, lines, areas, or pie slices:

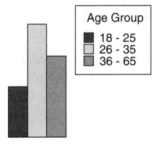

A liar might well use too-similar colors in the legend of a chart—attempting, for example, to blur the distinction between otherwise unlabeled plots of profit and loss.

Color as a Rendering Tool

The number of color tones in a photographic scene might number in the thousands—from a possible range in nature that might number in the millions. From the standpoint of the range of color choice, a fine artist would have the same requirement. You wouldn't want to restrict the color choices of a budding Picasso to a few simple codes. Use of color to reproduce a scene in nature or as artistic expression would require the widest possible range of colors (see Figure 11.2). Notice that simply reproducing the original picture as a monochrome photograph not only loses the richness of full color but also makes the scene lack depth.

Figure 11.2   *Rendering a scene from nature requires the widest possible selection of colors—the opposite of the requirement when color is used as coding. (Photo by Gary Palmatier.)*

The need for richness and flexibility in color selection is just the opposite from the requirement for color codes. Instead of restricting the number of colors, you need to have all the possible subtleties of hue, shading, and tint.

The world of product photography for advertising is full of liar's tricks. To heighten its visual appeal, food is often photographed through a pink filter, which conveys the feeling of warmth. Nobody wants a hamburger that looks slightly blue or green! In some cases, selective retouching will be required—even though you might add green to make that salad to look cool and crisp, you don't want its surroundings to look chilly.

Color as a Design Element

When you are designing a business presentation, the range of colors you need will fall somewhere between the extremes of natural color and color as coding.

In general, your basic approach will be to use colors as codes, as in keying lines and bars to their respective company divisions or product lines. For these chart elements, the rule about coding applies: Keep the number of codes to a minimum to avoid confusing the audience, and make sure that the color tones you use are distinctly different.

Color can also be used as *thematic* coding—to distinguish different sections of a longer presentation. For example, the marketing presentation might have a blue background, followed by the financial section in green. Although the range of colors here might be considerably wider than the basic codes you've used in charts, you should still try to limit the number of color themes. Too many color themes—which are, after all, a type of coding—will confuse the audience. An effective technique is to build each major section of a presentation around a basic color hue, then vary the tints or shades of that hue within the section.

Your overall objective in choosing colors should be to reinforce and clarify the content of your presentation. You should think of color as one of many essential design elements such as font, line weight, and spatial composition. Used in coordination, these elements can produce a visual composition that is readable, understandable, and aesthetically pleasing.

Color Range in Business Presentations

Because too many color codes or color themes can confuse your audience, there is no obvious need for a wide range of color in the presentation itself. Nevertheless, you will need a wide range of colors from which to choose—enough variety to permit you to select a few precise tones that work well together. When your needs move beyond coding to include pleasing and effective color composition, the range of colors needed actually becomes quite large. To understand why this is so, you must understand some of the factors that determine color range.

"How many colors do I need?"

The question might seem somewhat academic here. Isn't color selection mainly a matter of taste? Not at all. These days, most business presentations are created on computers. It's just too much work to assemble artwork by hand, and the charts won't be as accurate. Reproducing "full color" still involves a relatively expensive set of computer hardware options, and it's not uncommon to run into some severe constraints. If you have a computer that can display just 16 colors (and those are becoming rare), you might think you have more than you need. Have a look at the following section on color palettes to find out why you'll probably be changing your mind about how many colors are enough.

Color Scheme, Palette, and Gamut

Your color choices for a given visual composition make up the *color scheme* of the composition. Within a computer, the set of specific colors available is called a *color palette*. The range of colors that can be reproduced on a specific video monitor or printer is called its *color gamut* (see Figure 11.3).

Figure 11.3 *Here are some more precise terms for color selection.*

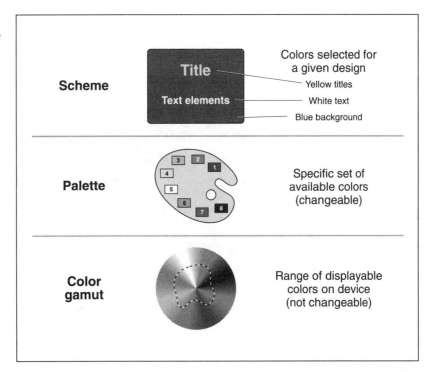

A color scheme is an abstract idea—your notion as a designer of how a presentation should look. A color palette is a tool for carrying out a color scheme. The color gamuts of computer devices are physical, technical characteristics that determine which color tones you can choose for your palette.

So, when you are choosing a palette for use with graphics software, you need to think about both your ideal color scheme as well as the color-gamut constraints of the computer. Your color selections, therefore, will be a compromise between your desired color scheme and the color gamut that is available to you.

Color Selection

As shown in Figure 11.4, your preliminary decisions about a color scheme for your charts should include selections for background, main title and subtitles, text (explanatory notes, lists, and chart labels), highlighted text, and subdued text. (For an example of build-and-subdue technique for text charts, see Figure 10.8 in the previous chapter.)

Figure 11.4   *Here's a set of typical color choices for the text charts of a business presentation.*

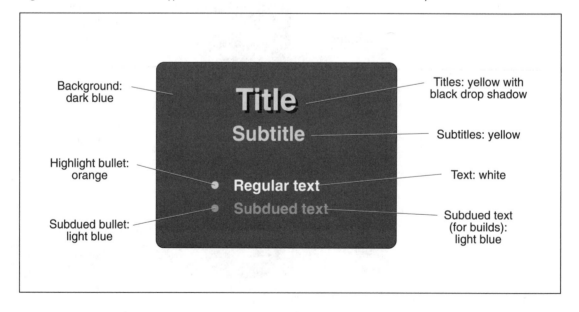

I can't emphasize it too strongly: *Background color will be your single most important choice.*

The background color you pick will largely determine the color scheme and the color palette you use to carry out that scheme. The specific tones in your color palette will be influenced by the type of output medium (such as paper or video screen), room lighting conditions, the mood you want to create, and whether your presentation will also be published in other media, such as on paper handouts.

Remember that the value of a color refers to its degree of darkness. The value of the background in a color scheme determines the *contrast range* of the entire picture—the range of values between the lightest and the darkest color tones. The greatest contrast generally should be between the background and the titles or text—because *contrast promotes readability.* Again, the controlling factors include the output medium, lighting conditions, mood, and reproduction requirements.

Examples of output media include color slides, overhead transparencies, paper handouts, and video. The most effective color slides have relatively dark backgrounds. But, you should increase the value of the background shade in proportion to the brightness of the light in the meeting room. Brighter room lights require lighter background colors and a narrower contrast range, as between background and text.

So, the colors you choose for titles and text should be among the lightest colors. You should pick subdued, or dimmed, color shades for toning-down or de-emphasizing certain lines of text or areas of a chart. An appropriate subdued color for slides would be a lighter shade of the background. Highlight color is used to draw attention to specific elements or objects. This color should be the brightest of the colors in the chosen color palette. Examples of the same chart designed for different media are shown in Figure 11.5.

For overhead transparencies, the opposite is true. The most effective overheads have light-colored backgrounds. While slides usually are viewed in darkened rooms, you might prefer to use overheads when the room lights must be left on, as in training sessions or demonstrations—or just when you want the audience to be able to see you

Figure 11.5 *Here are some rules of thumb for selecting colors for different presentation media.*

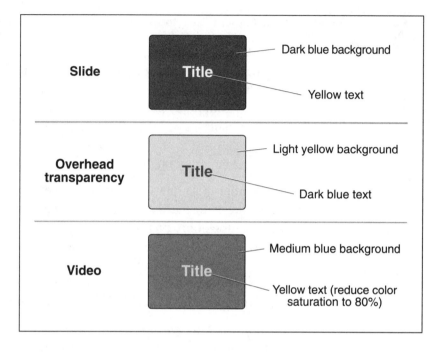

clearly. For overheads, the brighter the lights in the room, the deeper the color of the background must be. Titles and text are usually rendered in dark colors to provide sufficient contrast. Subduing and highlighting aren't usually done on overheads, since the range of usable color tones is much narrower when all of them must be light-shaded.

Build-and-subdue techniques are not usually done for overheads, since this might involve possible fumbling of the multiple frames when just one will do. Instead, the presenter can cover portions of text with a blank card (called a *slip card*), sliding the card downward to reveal each line or section of text as the discussion progresses.

Returning to the topic of contrast, graphics produced for video require relatively narrow contrast ranges because of the technical characteristics of video electronics. Slides that project well in dark rooms have too much contrast for successful transfer to videotape.

The mood or content of the presentation is another consideration in choosing a background color. A deep blue is by far the most popular color for the backgrounds of business slides. There are two reasons for this:

- Blue objects tend to recede, or stay in the background, in relation to other colors. Blue contrasts well and does not detract from other elements of the composition.

- Blue is the most popular color for corporate logos. This is a content issue: A company audience is likely to be biased in favor of their "school colors" as an overall theme for all slides in the presentation. Yellow and white for titles and text provide good contrast with blue.

Other effective background colors for slides are green and dark gray. Both of these colors are easy on the eyes and compose well with the colors of other elements.

At least for Western audiences, avoid color combinations that are traditional for holidays—red and green for Christmas, blue and white for Chanukah, orange and black for Halloween, yellow, pink, and pastel blue for Easter. This rule is analogous to avoiding familiar tunes in the sound tracks of recorded presentations: well-known color or music schemes will evoke feelings in your audience that have nothing to do with your message.

Pastel colors are often used for the backgrounds of overhead transparencies. Yellow, pink (or light magenta), light green, and light blue are good choices. Titles and text might be dark blue or black.

Backgrounds for video must be chosen to decrease the overall contrast range of the picture. The guidelines are much the same as for color slides, but the combined intensities of the value and chroma—or saturation—of video backgrounds should not be more than 80 percent of primary, or fully saturated, colors.

Compatibility with other output media should be your final consideration in choosing a color palette. Converting slides to video or slides to paper handouts, for example, usually involves some redesign for reproduction in a different medium. For conversion to video, the contrast range and color saturation of any charts designed for projection will have to be reduced.

In converting to paper, you might be limited to the *grayscale*—or range of monochrome shades—of a black-and-white printer. In such a printer, for example, the tones of red and green that you used might both translate to the same shade of gray. To differentiate the colors in the monochrome version, you might have to use patterns to supplement the shading.

If you plan to duplicate your visual materials, bear in mind that most forms of duplication, or copying, add contrast with each reproduction step—or generation. For example, slide originals designed for duplication must have narrower contrast ranges than slides intended for projection.

How Many Bytes Will It Take?

Color capability of computer graphics systems results from two technical considerations:

- How many levels of intensity can be produced by the color electron guns (or other recording instrument) in the display or printer?

- How many data bits are associated with each point, or picture element (pixel), in the image?

Both factors have hardware and software components, but their net effect is to limit the color variations that can be displayed. In the case of color monitors, the variations are among red, green, and blue electron guns of the color picture tube (called a *cathode ray tube*, or *CRT*). In the case of plotters or ink-jet printers, it is the number of pens or inks that can be controlled and the combinations and patterns with which they can be laid down on paper.

Again, color as a design consideration is subjective. The only valid generalization, even among the comparatively narrow ranks of business graphics users, is that you will always end up needing more than you thought when you acquired your computer system or set up your color palettes.

These requirements become important when you are evaluating computer systems for graphic applications. Color capability of displays is usually referred to as a subset of simultaneously displayable colors of some total set, or palette, of colors. Color ranges also depend on the processing capabilities (word length in bits) and memory capacities (in bytes) of the computer, video card, or printer.

Colors and Computers

You would think that people could agree on how many colors are enough. As I point out earlier in this chapter, process control engineers who design displays for monitoring nuclear reactors might tell you that four colors may be too few—and eight too many! Their assumption is that, as a scheme of coding, too many colors are apt to be confusing.

Indeed, some accountants feel this way about business slides: Too many colors are distracting and even somehow dishonest—an attempt to obscure the underlying meaning with dazzle. However, it's been my experience that businesspeople—perhaps because of the influence of advertising and the perception that "color sells"—expect color to be used liberally. Even if you take the conservative position that board-room graphics should be restricted to, say, eight colors, that begs the question: *Which* eight?

Evaluating the color capability of graphics systems—both software and hardware—is not a simple task. If you want to be able to discriminate among four shades of corporate blue, for example, you must also have that degree of discrimination among the other basic colors. Already, you're way beyond eight color tones.

Choosing Computer Colors

Presumably as an aid to the designer, most computer graphics programs offer predefined styles of presentations, which usually include color palettes. If you let some computer program control the design of your presentations, you might have no reason to learn any more about color selection. But just as soon as a supervisor or client requests a color change in a chart you've made or wants a different color scheme, you need to be concerned about details such as color palette definition, order of color assignment, changing of color attributes, correspondence between colors and monochrome textures, and so on.

There are three basic ways of selecting colors in most charting programs:

- Select a predefined color scheme. The color scheme is controlled by the order of colors in a color palette.

- Change the colors of objects explicitly by assigning them different colors from the palette.

- Mix *custom* colors, which can be used to replace colors in a palette or can be given to specific objects.

How Color Palettes Work

In computer software, a color palette is an ordered arrangement of a set number of colors (or shades of gray if it is a monochrome palette), as shown in Figure 11.6. (This example uses Freelance Graphics for Windows, but the palettes of other programs are very similar.)

In the example palette, positions of colors in a palette are numbered 1–64. This numbered order is the sequence in which the program assigns colors to the different types of graphic objects that you create. Specific assignments of palette color numbers to types of chart objects are shown in Table 11.2.

Figure 11.6   *In computer charting programs, palettes are tables of color-assignment numbers.*

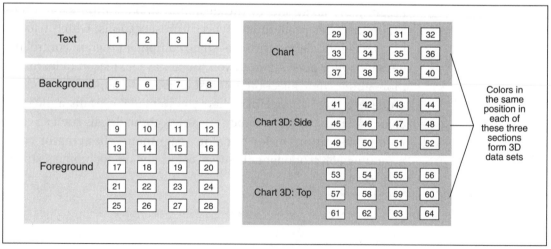

Table 11.2

Assignment of Palette Color Numbers to Object Types

Palette Number	Object Type
	TEXT
1	Titles
2	Subtitles, bulleted text, added text
3	Data chart text
4	Dimmed gray (subdued text)
	BACKGROUND
5	Solid background main color
6	Patterned background secondary color
7	Alternate background main color
8	Alternate background secondary color
	FOREGROUND
9–14	Drawn objects
15	Area and edge of drawn object
16	Chart frame edge
17	Chart frame area (graph background)
18	Organization chart box area or table background
19	3D organization chart box side
20	3D organization chart box bottom
21	Shadow
22	Bullet, line, area
23	Organization chart line
24	3D data chart floor
25	Chart and graph grid
26	Edge (outline in charts)
27	Table border, cell border
28	3D chart wall
	CHARTS
29–40	Data sets 1–12 (A–L)
	CHART 3D: SIDE
41–52	Data sets 1–12 (A–L)
	CHART 3D: TOP
53–64	Data sets 1–12 (A–L)

In effect, chart objects are normally colorless. Instead of having a specific color, each object has a color number. The object will take on the color or shade of gray found at that position in the current palette. Here, for example, is how the colors of 3D bars are assigned from a palette:

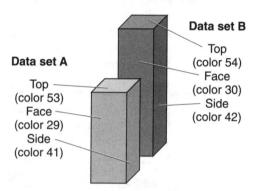

Simply by selecting another palette for use with a presentation, you can change its entire color scheme. Each object in the presentation will take on the color found at the same numbered position in the new palette.

It might be best to think of color palettes as sets of numbered color assignments. Remember, too, that the colors need not be all different. For example, for design purposes, you might want Color 1, used for titles, to be the same as subtitles (Color 2), chart frame edges (Color 16), bullets (Color 22), and table borders (Color 27).

If you change a color in a palette, you change the color of all objects in the presentation that have been assigned that color number. For example, if you change Color 1, which is usually assigned to chart titles, from yellow to white, all of the chart titles in the presentation will be changed from yellow to white.

For any color number within a palette, you can change its color assignment simply by selecting another one of the predefined colors. Or, you can create your own colors by mixing, using either the RGB or the HSL methods.

Mixing Custom Colors

You might think of palette colors as relative and custom colors as absolute. Objects that take their colors from a palette, such as chart elements, will change color when you change palettes. Objects to which you have assigned custom colors, such as drawings, will stay the same even if you change color palettes.

In technical terms, objects that have custom colors are assigned explicit RGB values rather than color palette numbers.

In practice, you might never need to mix custom colors. But you might need to do so if you need to create custom palettes to match the color reproduction characteristics of a specific printer, for instance.

Some computer graphics programs use the RGB method of color mixing. That is, you mix a custom color by making separate adjustments to its red, green, and blue components.

Mixing Colors by the RGB Method

To mix a custom color by the RGB method, you must make separate adjustments for Red, Green, and Blue. The adjustments typically range from 0 through 255 for each of these primary colors. In charting programs, you usually change the settings by typing numbers for the values or by adjusting graphic controls such as sliders (which work like volume controls on a stereo).

If you have a color in mind, it won't be easy to get it by mixing red, green, and blue. The best use of the RGB method is to adjust numeric settings to match values in published tables for printers and other output devices.

Although mixing RGB colors by eye can be cumbersome, here's how to do it if you must: To change hues, add one or two of the component colors. For example, adding some red and some green will increase the yellow in the color you're mixing. Adding blue and green will produce blue-green. Adding blue and red will produce violet.

Increasing red, green, and blue equally will have the effect of lightening, or adding white to, the color. Decreasing them equally will darken the color.

Mixing Colors by the HSL Method

It is easier to mix colors by eye using the HSL method, where you adjust the hue (position in the color spectrum), saturation (tint, or intensity), and lightness (value, or shade). Like the RGB method, you make adjustments by typing numeric values or by adjusting graphic controls.

Changing the Appearance of a Presentation

In most charting programs, you don't control the selection of the color palette directly. Instead, you pick a "look" for the presentation, which is based on a predefined style—a collection of options that includes a color palette.

In computer terms, a style is a collection of *templates*—usually, one for each chart type. You can think of a chart template as a generic description of a particular type of chart, such as a bar graph, which requires only that you supply the data.

Chart templates, in turn, are collections of options. These options include page layouts, as well as object attributes such as edge (outline) colors and fill (solid area) colors, and line styles (solid, dashed, or dotted). When you create a new presentation, the options in the template you choose are applied automatically to all the charts you make.

At some later time, after you have created a presentation, other styles and palettes can be applied to reset many options at, literally, a stroke—much as the appearance of word-processing documents can be changed by applying different templates, or *style sheets.*

Viewing Colors on a Computer Screen

The actual colors you see on the screen of a monitor can vary considerably from one computer to another. For example, if you have a computer that can display 16 colors, only 16 of the palette colors (including black and white) will be shown as solid, or "pure," colors (see Table 11.3). The others will be reproduced by an approximation

Table 11.3
Pure Colors on 16-Color Computer Displays

Color Name	RGB Values		
	RED	GREEN	BLUE
Red	255	000	000
Yellow	255	255	000
Olive	129	129	000
Neon Green	000	255	000
Dark Green	000	128	000
Turquoise	000	255	255
Aztec Blue	000	130	128
Flag Blue	000	000	255
Midnight	000	000	128
Hot Pink	255	000	255
Plum Red	128	000	128
Scarlet	129	000	000
White	255	255	255
25% Gray	192	192	192
50% Gray	128	128	128
Black	000	000	000

called *dithering*. Dithering creates a dot pattern that uses two colors to produce the impression of a third. The number of solid colors in the gamuts of other computers are 256, 32,000, or 16 million—quite a range! The more colors in the gamut, the more will be displayed as solid rather than dithered.

Even if colors appear dithered on the screen, they can be reproduced as solid tones on some high-quality output devices. For example, *color film recorders* are printers, in effect, for generating photographic slides as computer output. Most color film recorders can reproduce 16 million colors.

Those Sexy Chart Backgrounds

A chart background can include all kinds of graphic elements, including not only color, but also drawings or pictures on which charts and text can be overlaid. A background can be a solid color, a patterned blend of two colors, or even a picture. Optional elements would be decorative objects such as borders, symbols, and logos, which often are incorporated in predefined presentation styles (templates).

Liars are fond of using pretty pictures—showing attractive people, showy vehicles, or exotic vacation resorts—as backgrounds for less stimulating, but perhaps much more significant, material. Consider, for example, the timeshare vacation presentation that shows a sunset picture of Puerto Vallarta behind a chart that is supposed to disclose how your condominium investment is expected to appreciate. The audience might be forgiven for remembering more about the sailboats than the sales data!

For best results, the first and second colors of blended backgrounds should be contrasting values of the same color hue, such as light blue and dark blue. When you mix two different colors in a background, some of the gradations of blended color either won't be pleasing to the eye or will be so garish as to take attention away from the objects in the foreground.

How Do You Like This Picture?

Computers typically handle background patterns and pictures as *bitmaps*—arrays of tiny, colored dots. Even though patterns and bitmaps can make attractive backgrounds in color slides, you may wish to avoid these effects. Some output devices, including most printers, cannot reproduce them or cannot reproduce them well. In most cases, printing time can be slowed greatly. Also, when used in large areas, these effects add significantly to the size of the computer disk file in which the presentation is stored.

Changing Background Color

In a charting program, one way to change the background color for all charts in a presentation is to edit the color palette (have another look at Figure 11.6). To change a solid background in the example palette, you would change Color 5 (the first color in the second row). To change a patterned background, you would change both Colors 5 and 6.

Designing for Monochrome Output

A straightforward way to design color presentations for black-and-white or monochrome output—such as paper handouts—is to switch to a palette that has only shades of gray. The program might do this automatically when you select a monochrome output device for printing, but you won't have much control over the result.

Designing charts in color for grayscale output is rather like painting pictures for a person who is color-blind. Remember that you can have very different colors that produce the same grayscale value on paper. Red and green areas of equal intensity (RGB number) will have the same grayscale value and, therefore, will look identical in monochrome. Furthermore, the blue component in any color will be *almost entirely ignored* in the monochrome version.

Liars really don't care whether their color presentations are reproduced well in the black-and-white handouts they leave behind. (That's assuming they have the courtesy to give you any handouts at all!) A really nasty trick would be to show a critical plot in light blue—which will disappear completely when printed in grayscale!

The best solution for presentations that will be reproduced in other media is to use separate color palettes—a full-color palette for slides and a grayscale palette for monochrome printing on paper. You should adjust the shades of the grayscale palette so that there is as much contrast as possible between plots of data sets, for example. If manipulating the grays doesn't result in charts that are absolutely clear, use different patterns, as well.

Designing for Printed Color Output

You might think that a solution to reproducing your color slides faithfully on paper would be to use a color printer. In fact, you must do as much adjusting for color paper output as for monochrome.

Video screens and color film recorders use *additive* color mixing. The result of adding red, green, and blue—all at full intensity—is white.

Printing on paper and exposure of photographic films rely on a process of *subtractive* color, based on how light is reflected through colored dyes, such as printing inks. The RGB and HSL color mixing methods can't be used directly with subtractive colors. When you design presentations for color printing, you must use the *CMY* method, named for its component colors—cyan, magenta, and yellow. Color printing involves layering patterns of ink in these colors on white paper, which reflects light back through them to create color mixtures. However, the combination of cyan, magenta, and yellow at full intensity does not produce a true black (as RGB does with each component at zero intensity). To make up for this, full-color printing uses a fourth layer—black, designated *K* to avoid confusion with blue. The whole scheme is called *four-color process,* or *CMYK.*

You don't have to worry about the technicalities of translating among RGB, HSL, and CMYK. Many charting programs will do this for you. Another guide is the Pantone Matching System, or just *PMS Colors*. A long-standing vendor of specialty art papers and printing inks, Pantone, Inc., has developed a system that establishes equivalents between standardized color names (such as "Sangria" for a particular tone of red-violet) and their different numeric descriptions in RGB, HSL, and CMYK.

Beyond the matching of colors, be sensitive to differences in color-palette requirements for putting charts on paper. In general, the rules for designing overhead transparencies will be more appropriate than those for slides: Use light-shaded, neutral, or pastel backgrounds and dark text.

For Extra Credit: Using the Color Wheel

Yes, choosing attractive color combinations for your charts is largely a matter of taste. But by now you should be aware of many other important factors that, at the very least, might overrule your selections.

A device for making attractive color choices is the *color wheel*, shown in Figure 11.7, which you may recognize from art class in elementary school.

Figure 11.7 *Remember the color wheel from art class?*

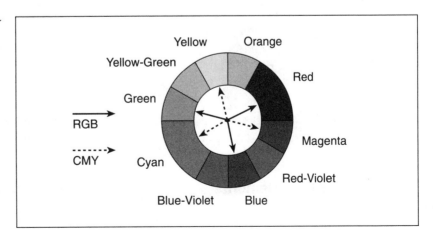

The color wheel shows how the primary hues of red, green, and blue can be blended to produce the other hues. The blends of the primary hues are called *secondary* hues, such as yellow-orange. (Note that the primary colors of the RGB color wheel are the secondary colors of the CMY wheel, and vice versa.)

According to color theory, pleasing combinations of hues result from certain positional relationships on the wheel, as shown in Figure 11.8.

Figure 11.8 *Positional relationships of colors on the wheel produce combinations that are pleasing to the eye.*

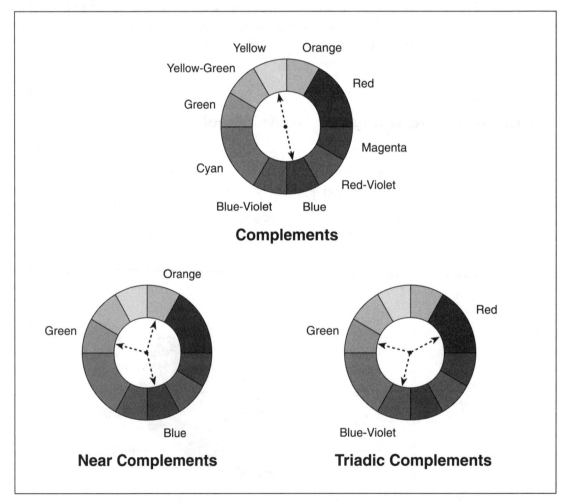

By color-wheel position, there are three types of color combinations: complementary (opposite on the wheel, shown on top), near complements (left), and triadic complements (right).

Each of the predefined color palettes in charting programs is set up according to one of these three methods, based on the main selection of the chart background color. If you don't have a ready-made palette to choose from, you now know what the professional designers know about putting colors together!

Getting Smart About Color—A Case Study

In one of my previous career incarnations, I was creative director for a company that produced business presentations. I offer the following true story as a cautionary tale about the uses and abuses of color.

My assignment was to produce a slide show for an audiovisual agency in Detroit. The agency's client was an automobile manufacturer—an influential customer, especially in that town. The slide presentation included word slides, charts, and, of course, stunning photographs of the new car model.

For the computer-generated slide graphics, the client chose backgrounds of deep burgundy. When the original slides were projected, they were gorgeous, in my humble opinion. The burgundy conveyed richness and luxury. I was sure that the client would be pleased.

The agency, which had contracted elsewhere for the car photos, made the decision to handle its own slide duplication through one of the professional film laboratories.

When the slide copies came back from the lab, I received an angry call from the agency. The beautiful burgundy backgrounds had turned an unattractive brown—and the film laboratory insisted there was nothing to be done!

I had another look at both the originals and the copies, then paid a visit to the film laboratory. I knew that the overall tint of the film copies could be adjusted to more closely match the computer-generated colors. I had a long and heated conversation with the technical chief, who explained, "People in this town look at three things in a film print—people, sky, and cars. When we adjust the tint, we go first for the faces. And auto paint is very difficult to reproduce, especially the metallics. We have to get the paint job to look just right without turning the people green. The graphics just aren't my worry. Who else but you knows what color they were to start with?"

The situation had been made worse by the selection of the exceptionally dark burgundy background color, which was nice for projection but could not be rendered properly by the duplicating film without sacrificing some other range of colors in the rest of the presentation. Unfortunately, that hearty burgundy now looked more like burnt pot roast. But, if we were to adjust the tint of the copies to enhance the background color, we'd ruin the live photography, including the color of the client's product—that shiny, new car.

Through my previous long and painful experience dealing with the auto makers, I had learned the consequences of getting the car color even slightly wrong. When one of the executives saw some film prints I had adjusted for color, he demanded, "Where did you get that vehicle? We don't even make that color!"

The solution to what I'll call "The Case of the Muddy Burgundy" was to make a special set of slide originals—with lighter backgrounds in the same hue—for duplication.

So, consider that a single photograph of a client's product in a slide show could very well dictate the colors of the entire presentation—that is, if you aim to show the truth, as the client sees it.

And that's not just a matter of taste!

How Not to Get Cheated On

THIS concluding chapter recaps the tell-tale clues that a chartmaker is cheating. Watch for these tricks, and you won't be fooled! As I've said before, what you do with this knowledge should be a matter of personal conscience. (Remember that they allow you one phone call, and my number is unlisted.)

If you want to be like Hamlet's crab and go through life backwards, you could productively start reading this book here, delving into the previous material as you find potentially useful topics.

Chapter 1: The Numbers Don't Lie—*Do They?*

Labeling can make the difference between having raw data and having useful information. Leaving a label off an important number is a favorite trick of obfuscators and graphic deceivers. Worse, using a label that is only slightly incorrect is the mark of a truly skilled liar.

19 WHAT?

Getting to the infamous "bottom line" is one type of summary, or data reduction. Summaries can make raw data more digestible, but you lose all the flavors of the underlying detail.

Like other kinds of data reduction, arithmetic averages sacrifice detail in favor of a single, more understandable number. And just as relying solely on the bottom line is unwise, basing conclusions exclusively on an average can lead you to incorrect conclusions.

$$\frac{79 + 81 + 78 + 80 + 76 \text{ STROKES}}{5 \text{ GAMES}} = 78.8 \text{ AVERAGE STROKES PER GAME}$$

There are several other techniques for generalizing about a set of data, all of which might seem much the same to the amateur chartmaker and to unwary audiences. Liars can exploit your lack of discrimination by using these terms interchangeably: mean (mid-point), median, and moving average.

These other forms of data reduction have been developed because averages don't always represent typical values. A true average will be more reliable if: 1) there is a large number of individual data values, thus minimizing the effect of any one of them on the result, and 2) the data values are fairly consistent, or without wide variations, thus reducing the effects of unusual values on the result.

Just as dangerous as data reduction is any attempt to *expand* a set of data, either by guessing in-between values *(interpolation)* or by identifying trends and projecting future values *(extrapolation)*:

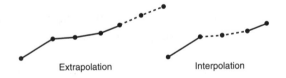

Extrapolation   Interpolation

The lesson of data reduction is that, as a conscientious chartmaker, you can never avoid showing your own interpretation of the data. The trick, if you'd call it that, is to make a chart's interpretation of the data support and reinforce your business message.

Chapter 2: Pies

The familiar pie chart is the most overused, misused, and sometimes downright useless trick in the presenter's repertoire. You can use a pie chart to show any proportional relationship between a slice and the whole pie—as a percentage (20%), a fraction (1/5), a ratio (1:5), or a decimal (0.20). The trouble starts when you begin to think of a pie in terms of *amounts*:

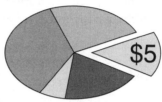

Resist the temptation to put values instead of percentages on the slices of a pie chart. If the actual values are essential to your business message, use another type of chart.

Avoid showing two or more pies in the same chart. The almost irresistible temptation will be to size the pies differently, presumably to reflect their different total amounts. This is wrong for two reasons: 1) You will fall immediately into the trap of showing amounts rather than percentages, and 2) Your intuitive notions about how big to make the pies will probably result in exaggerating the importance of the larger ones.

In a 3D pie, people will usually think that the *bottom slice* is the most important. That's because the dimensional effect distorts the apparent size of the slice by literally giving it an edge. People will perceive this slice as bigger than it would appear in a circular, two-dimensional pie:

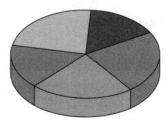

Liars who want to diminish whatever the pie represents will often clutter the chart with lots of labels, forcing the pie to be smaller to fit on the page or screen.

Liars sometimes find it inconvenient that all pieces of the pie must be accounted for. They typically get around this by labeling a mystery slice All Others. A related trick is to leave a troublesome part of the data out of the pie, so that the whole isn't really whole.

There is one situation in which I recommend that you cheat in preparing a pie chart. If you must include a very small slice—equivalent to 1 percent or less—plot it at about 1.5 percent (about 5 degrees of the circle) so that the thin slice can be seen. Be sure to add a descriptive label nearby that discloses the true percentage:

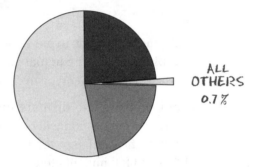

Alternatives to multiple pie charts are donut charts and 100-percent bars. Be cautious about using these, though, if your audience is unfamiliar with these charting conventions. If you intended to show amounts rather than percentages, consider using stacked bars or stacked areas.

Chapter 3: Orientation

Viewers get very different impressions depending on whether shapes appear to go left, right, up, or down. Rightward motion—the way the eye scans a page—is associated with the passage of time—and so with positive movement, and even the idea of progress. Motion upward means gain, and downward means loss.

These are the basic assumptions behind the Cartesian coordinate system and the *xy* chart: Time flows from left to right, and the amount of the thing being measured and charted fluctuates up and down.

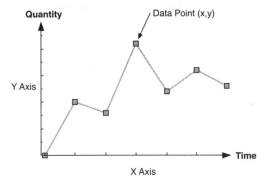

When your audience shares a set of biases, a presentation will be more effective if its design plays to those assumptions deliberately. Don't try to neutralize their biases—which probably isn't possible anyway. Simply make your designs consistent with what they expect.

Linguistic and graphic conventions about left-right and up-down carry over to concepts about geography, which we get from relying on two-dimensional maps. Particularly for Westerners, up-down-left-right in the screen can be interpreted as north-south-east-west on the globe. These concepts should also be considered in your graphic designs—including the composition of labels in charts, photographs of plant locations, scenic chart backgrounds, tables of demographic data, and diagrams of computer networks:

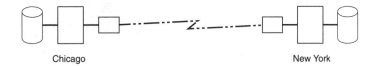

Chapter 4: XY Charts

In an *xy* chart that plots quantity versus time, amounts increase or decrease along a vertical scale, or *y axis*. Time progresses from left to right along a horizontal scale, or *x axis*.

In the most common type of *xy* data chart used in business, a sequence of data values is plotted as a series of points on the chart in left-to-right order. The height of each point is its *y* value. Its *x* value is the next division of the *x*-axis scale:

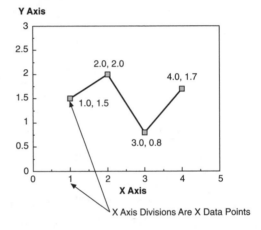

The sequence of points is called a *data series,* or *data set.* Sets of data can be built on one another in two different ways: stacked or cumulative.

In a stacked area or bar chart, each data set uses the previous set as its baseline. A liar can exploit this characteristic of stacked charts. The fluctuations, or bumps, in the lower data series will enhance and exaggerate the bumps in the upper ones. The honest alternative is to put the series with the least variation on the bottom:

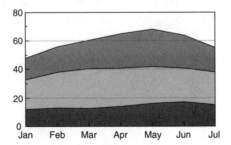

The bumpy effect will be most pronounced in area charts but can also affect perceptions about stacked bar charts.

A liar might also try to stack the lines in a line chart—but this would be a grievous error. The notion of stacking won't be visually apparent, as it is with solid areas or bars. By convention, the baseline of each series in a line chart is usually assumed to be the line $y=0$ (the x axis).

For bar charts only, an alternative to stacking several data series on the same chart is to cluster them:

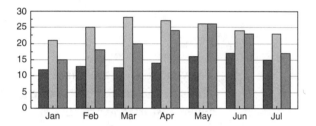

In a cumulative chart, each point within a data series is a cumulative total of all preceding points in the same series. Cumulative charts are a favorite with liars because the general impression is always optimistic: As long as the data points are positive numbers, the trend is always upward—even if some of the increases are small.

Don't confuse *stacked* and *cumulative*. Stacking data series does make them appear to accumulate on top of one another, but this is not the meaning of *cumulative*.

High-Low-Close-Open (HLCO) charts, a special *xy* format for showing stock prices, can be distorted by playing with the appearance of the bars (skinny or fat) and by manipulating the *x* and *y* axes.

In the Gantt charts used for project management, durations of work activities are plotted as horizontal bars. Project managers abuse Gantt charts this way: If the *x* axis is elongated, the audience will get the impression that the overall time span of project tasks has been lengthened.

Two of the liars' favorite tricks with *xy* charts are: First, turn vertical charts into horizontal format, subverting the intuitions of the audience about quantity (up-down) and time (left-right):

For your second trick, mix proportional and quantity-time relationships in the same chart. For example, use symbols instead of bars, which distorts their relative sizes:

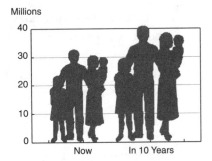

A favorite trick of liars is to use 3D effects to make all the bars in the chart appear taller. The distortion will be most pronounced by a trick of perspective that places the vanishing point above the tops of the bars:

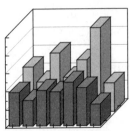

Even if the *x* axis does not represent the flow of time, viewers will regard the bars on the right as the most recent or important. Even if the *x* axis has nothing to do with time, the audience won't be able to avoid associating left-to-right movement with progress. Liars therefore will put the set of bars they want to emphasize on the right end of the chart (but even honest people can do this trick with a clear conscience).

Chapter 5: Radar Charts

In a radar chart, several axes radiate outward from a common center like the spokes of a wheel. The optional grid lines are shown as concentric circles, each one a uniform distance from the center. A data point is plotted at a measured distance from the center on a grid line. The data point is typically a score or rating for the criterion or category represented by the axis, or spoke.

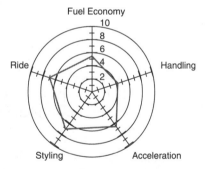

The radar-chart format imposes a bias on the interpretation of the data: The data set with the most nearly regular shape will be the most well-balanced among the various criteria. Symmetrical plots are therefore "good," and asymmetrical "bad." This bias is just as strong—although decidedly different from—the up-down, left-right biases inherent in *xy* charts.

If you want to be able to draw conclusions from the shapes of the data plots in a radar chart, the scoring must be consistent for all criteria. Furthermore, you should convert all sets of scores to the number of decimal places in the least precise scale. Cheaters might pretend that the dissimilar scales on their radar charts are proportional to each other:

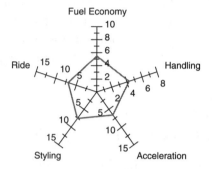

A particularly nasty trick with radar charts would be to intermix valid charts with questionable ones as adjacent pages or slides in the same presentation.

Chapter 6: Axes

On an *xy* chart, an axis is one of two numeric scales on which the chartmaker plots data points. Axes also help readers of a chart estimate the numeric values of its plotted lines, bars, or areas.

To flatten a plot—minimizing its fluctuations—*increase* the *y*-axis range. To emphasize a plot—maximizing its fluctuations—*decrease* the *y*-axis range. The effect can be heightened if the high point of the plot extends beyond the maximum value on the scale, which unnecessarily exaggerates the magnitude of that data point.

The most truthful version of any chart will be the one that has the least deliberate distortion.

If it suits your lying purposes, don't label the starting point of the *y*-axis scale. Or, don't label any of the scale divisions except the top one. The wider the range and the more divisions it has, the easier it will be to trick the audience into thinking that the scale starts at zero. The honest alternative is to begin the scale at zero but indicate a break in the scale:

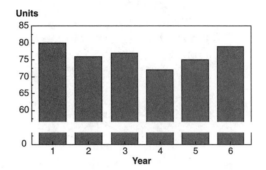

Scaling or using a multiplier on scale values can be a tool for deceitful manipulation of axes. Also, the magnitude of the data can also be either emphasized or minimized by increasing or decreasing the number of scale divisions and their corresponding labels.

Plots of different magnitudes can be shown on the same chart by using two y-axes, with one scale on the left and a different scale on the right. Liars know that such dual-*y* charts can be difficult to interpret if more than two lines or sets of bars are shown—one for each *y* axis. The damage can be compounded if there is no legend, or at least some obvious coding scheme for matching the plots to their respective axes. A clever liar might also use dual-*y* plotting to make bogus comparisons between performance data and some unrelated trend line:

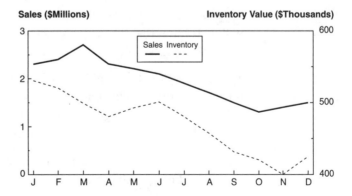

Logarithmic axes, or log axes, are scaled by powers of 10 (orders of magnitude). The legitimate purpose of log scales is to fit a very wide-ranging data set on a single, compact chart. An exponential trend can be made to look less dramatic by plotting it on a semi-log chart, where it resembles a linear trend:

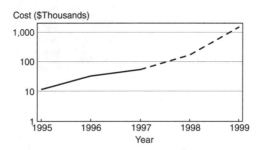

Chapter 7: Choosing Chart Types and Styles

Liars will choose chart types and styles that hide or minimize messages that might otherwise be obvious in the data. That is, liars will deliberately pick the wrong tool for the job.

A liar might show you a pie chart, but talk instead about the specific amounts of the total pie or of the slices, drawing the wrong conclusions. In a proportional format such as pie charts, the presenter should stick to percentages and avoid discussing amounts.

Another pie-chart liar's trick is to omit miscellaneous slices, which isn't immediately apparent because the audience will assume that the pie represents 100 percent of the data. Omitting inconvenient slices will also have the effect of increasing the apparent size of all the remaining slices.

The liar's trick with areas is a fairly simple deceit: Areas simply look more substantial than line plots. Because an area emphasizes total volume, the audience might be less concerned with the peaks and valleys than if the data was shown simply as a line.

Liars also tend to mix stacked and cumulative techniques in the same chart—without making mention of the cumulative part.

Another liar's trick is to show things that depend on magnitudes—such as sales—as horizontal bars. As a liar might use it, the paired-bar format is an especially confusing use of horizontal orientation:

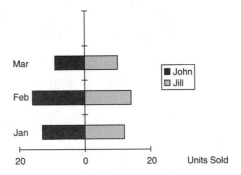

Save those lazy bars for showing time spans, such as the duration of tasks in a Gantt chart:

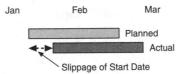

Venn diagrams are for showing concepts, not quantities. The sizes of the circles can relate to the relative sizes of the groups, but more often they must be sized to permit larger circles to contain smaller circles, showing relationships between subsets and the larger sets that contain them:

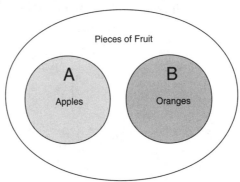

Chapter 8: Trends

Line charts can highlight trends in data better than any other chart type. They are used so often to show trends that audiences are prone to view all line plots this way.

Mathematical formulas can be used to describe trends. Formulas for extrapolation or interpolation work well if the underlying formulas accurately describe the real forces that are shaping the events we observe—especially if the process being described is fairly simple.

Recall from Chapter 1 that liars tend to confuse the important distinctions between *average, mean, median,* and *moving average.*

Error bars can be used to show deviations, or ranges of possible data values. Liars are not usually honest enough to even think of using error bars, much less make the distinction between *population* and *sample* standard deviation.

Regression is another trick of the statistician's trade for finding patterns in data. Common regression types are linear, exponential, logarithmic, and power. By trying different regression formulas, the statistician can fit smooth lines neatly through seemingly erratic sets of data points. The assumption is the better the fit, the more likely that the formula actually describes the process that is generating the data. Regression analysis can be tempting for sincere forecasters and liars alike because it produces smooth line plots and curves that can appear to fit the data neatly.

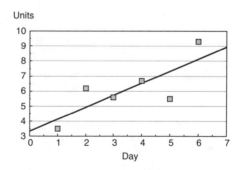

The pitfall with all kinds of trends is to oversimplify the data.

Chapter 9: Tables and Spreadsheets

How you set up a table in a report or presentation can determine whether it will inform or confuse your audience. Furthermore, if you use a table as a worksheet for chart data, you will find that the arrangement of its columns and rows strongly affects your ability to generate meaningful graphs from the data.

Liars might prefer tables instead of charts. An array of numbers in a table can be difficult to read, and it will be almost impossible for an audience to spot patterns in the data quickly.

 As a general rule, you should use a graph instead of a table to present numeric results. Tables are better as worksheets than as formats for presentation.

Even though accounting-style cross-footing is not necessary in electronic spreadsheets, many people still expect to see totals in one of two places—at the right end of rows or at the bottom of columns. So, presenting a table that has an unconventional layout, such as showing the totals at the top, can be disorienting to an audience.

Intermixing data categories in the same column of a table—making it difficult to compare entries in different rows—can be another disorienting trick of the accomplished liar.

In a report narrative or in a speech, you can avoid lengthy explanations of key relationships by summarizing them in a table. If the source data must be shown, consider presenting the numbers in a table beneath the graph:

	Aardvark	Aero	Acme	Able	Archer	Argo	Ace
Last Year ■	93	71	99	94	121	105	60
This Year □	124	83	162	95	143	231	78

Use bullets when presenting topics and key points, especially if you are listing topics that will follow as subtitled sections of your material. Use a numbered list instead if you a presenting a sequence of steps. You can usually increase the size and therefore the legibility of the text in a table if you use fewer columns, perhaps placing explanatory text in footnotes.

Instead of breaking a table, consider placing it on a separate page, even if it must be located farther away from its explanatory text.

Align alphabetic text, such as labels, on the left. Align numeric values on the right, usually with the right-most digit. If numeric values are decimals, align them on the decimal points (or use the same number of places for each string of digits, and right-align all the strings). Optionally, center the headings in their columns:

Company	Index
Acme	253.07
Arrow	4,841.372

The most important consideration in making a chart from a spreadsheet has nothing to do with computer software: You must exercise some careful judgment in selecting which data will be charted. In any sheet, there is almost always too much information to make a single, understandable chart. Out of the whole sheet, you usually need to find just one or two rows of data that tell a story.

Here are my Eight Great Steps to spreadsheet reliability:

1. Pick the right function for the job: ROUND or TRUNCate?
2. Watch for the kinds of errors the computer can't catch.
3. Make sure to give a function everything it requires (syntax and parameters).
4. Verify all references to data (addresses and range names).
5. Double-check your math.
6. Ask: Do these results *look* reasonable?
7. If you make changes to the sheet, recheck and retest it.
8. If you think your computer goofed, think again!

Chapter 10: Words for Your Charts

Inadvertently leaving off helpful labels on a chart is the quickest way to perjure yourself. Categories of things can be indicated by labels, which are as necessary to reliable data as the numbers, or values.

As a rule, wherever you see a number, you should find a label nearby.

Labels on data can sometimes be hidden or implied. Certain conventions for displaying results, such as delimiters and implied field names, are shortcuts to labeling data:

CALL (510) 523-8233
TO ORDER MORE COPIES
OF THIS BOOK!

It's a liar's trick to mislabel data, in effect, just by using delimiting conventions that are either unfamiliar to the audience or incorrect in the context of the presentation:

06.09.99
MEANS SEPTEMBER 6, 1999, IN EUROPE

In general, you should use data labels to: 1) Identify a specific data point you want to highlight in your presentation, such as an exceptionally high or low value, or an average, 2) Make sure that the audience interprets the correct value on plots that are difficult to read, such as estimating the height of 3D bars in relation to the chart scale, 3) Show the actual value for a small bar segment or pie slice that had to be made bigger just to be visible, 4) In an *xy* scatter chart, identify actual values from which trend lines were calculated (such as by extrapolation, interpolation, or regression), and 5) If the design does not permit locating the label close to the data point, use a callout line to direct the eye.

Here are some guidelines for labeling chart axes: 1) Include both the type of measurement and the unit of measure in an axis title. 2) You can omit an axis title only if it would be obvious from the scale labels. 3) If a scaling factor, or multiplier, is needed to interpret the scale labels properly, it must be included either in the axis title or in a footnote to the chart. 4) If an *xy* chart has dual-*y* axes, you must show an axis title for each. Ideally, the axis titles and legend entries should be color-coded to their respective data series.

Here are some other tips for labeling axes for better readability: 1) Use a skip factor to omit labels and regular intervals from a sequence. For example, if the skip factor were five, every fifth label would be printed on the scale. 2) If after using a skip factor the labels still seem too cluttered, adjust their orientation. The normal orientation of labels is horizontal. Some of the alternatives are: vertical up (not recommended), vertical down, slant up, slant down, and stagger.

It's a liar's trick to use color codes that are too similar in legends, making it difficult to correlate the names of data series with their plots.

Use footnotes in charts and in written reports to identify sources of information. The liar's trick with footnotes is pretty obvious: They don't use them!

Font attributes are ways of varying the appearance of a typeface, which can be done to enhance or clarify the labeling in a chart. Attributes can be applied to individual characters, or letters—or to groups of letters, such as words or even entire paragraphs or documents. Also, attributes such as bold and underline can be used instead of color in black-and-white presentations to highlight labels and emphasize explanatory text. Examples of font attributes include bold, *italic*, underline, double underline, strikethrough, color, hidden, small caps, all caps, spacing, superscript, subscript, and drop shadow.

When selecting fonts for charts: 1) Use a proportionally spaced font. 2) Use sans serif fonts for chart labels, and serif fonts for titles and explanatory text. 3) Don't intermix fonts within a chart. 4) Use upper- and lowercase letters.

For clarity, use correct punctuation and grammatical style in labels and explanatory text. In your report narrative, be sensitive to the use of pronouns.

Liars are prone to use the inclusive *we* in their speeches and reports as a way of associating themselves with known experts or reliable authorities.

When you are writing your report: 1) Avoid words having more than three syllables. 2) Keep sentences to less than 20 words. 3) Don't include more than two clauses in most sentences.

Professional speechwriters use title screens called *signposts*, which list topics to reinforce the structure of your presentation. Build-up sequences can be used to avoid the "read-ahead" problem when presenting lists:

Subjects of My Paintings
In Southern France

- Landscapes
- Still life

Subjects of My Paintings
In Southern France

- Landscapes
- Still life
- Live models

Make sure that the syntax, or word order, of the spoken words matches the text on the screen. A liar might not bother to signpost a presentation, or might do the opposite—list topics that are never discussed in detail.

Think of your speech or report as a long caption for your charts. If you say something that doesn't have a picture to go with it, come up with a new picture—or get rid of the material.

 **LIAR'S TRICK**

Liars know a trick we should all imitate: *Make all your mistakes in a loud, clear voice!*

Chapter 11: Color

Biases about color have to do with a person's professional and cultural background. But notions about color aren't that simple: Our interpretation of colors can change with the situation. People make these mental shifts frequently and without conscious effort, usually depending on their understanding of the purpose of a visual message.

Color can be used for 1) a kind of coding, 2) the rendering of real-life pictures and artistic expression, or 3) as a design element of graphic materials, including charts. Business presentations often include all three types of uses.

When colors are used as codes, the color scheme must be as simple as possible, with the least number of colors necessary. A liar might well use too-similar colors in the legend of a chart—attempting, for

Colors Are More Than Just Pretty

To an accountant, green money is good and red ink is bad. Doctors want it just the other way around: oxygen-rich red blood is very good, green necrosis very, very bad! But the situation is also important. The accountants and the doctors will agree on what to do when they see traffic lights on the drive home.

Liars are skilled at sensing the mood of an audience and then manipulating it to their own ends. A keen sense of how colors affect moods is therefore an essential skill for designers of presentations. Yes, perhaps using corporate blue is usually a safe plan. But it's just possible that using a lot of red in making an investment pitch to a group of doctors might stimulate more excitement than alarm.

example, to blur the distinction between otherwise unlabeled plots of profit and loss.

The need for richness and flexibility in color selection when you are showing product photography is just the opposite from the requirement for color codes. Instead of restricting the number of colors, you need to have all the possible subtleties of hue, shading, and tint.

Liars really don't care whether their color presentations are reproduced well in the black-and-white handouts they leave behind. (That's assuming they have the courtesy to give you any handouts at all!) A really nasty trick would be to show a critical plot in light blue—which will disappear completely when printed in grayscale.

When you are designing a business presentation, the range of colors you need will fall somewhere between the extremes of natural color and color as coding. Your preliminary decisions about a color scheme for your charts should include selections for background, main title and subtitles, text (explanatory notes, lists, and chart labels), highlighted text, and subdued text. Background color will be your single, most-important choice.

Liars are fond of using pretty pictures—showing attractive people, flashy vehicles, or exotic vacation resorts—as backgrounds for less stimulating, but perhaps much more significant, material.

Go Forth and Multiply...

...And add, subtract, divide—do all the mathematical magic you can conjure. I've spilled my guts, and now you know what I know, which might be just enough to get you into trouble! If you dare to reveal your numbers in a chart, remember that you can no longer pretend to be one of the uninitiated. If your trends are too trendy, your bars bizarre, or your orientation askew, the prosecution will nail you if the jury finds out that you read this book!

I'll say it again: Learning how to lie with charts will help you spot deceptive tricks, whether in someone else's charts or in your own. Where you go from here is up to you!

Unexpurgated Glossary

100-percent bar A proportional plotting style (an alternative to multiple pies) in which data series are shown as stacked bars, each having the same height, and each representing 100 percent of the series total.

A

absolute A value expressed as an explicit amount in some unit of measure; in mathematics, a value expressed as its positive distance from zero.

additive color mixing The result of adding red, green, and blue to render color tones on a video monitor or in a color film recorder.

address In a spreadsheet, the column letter and row number of a specific *cell,* or data storage location: A1 is the first (upper-left) cell in a sheet.

amount An explicit value of a data item, expressed as a number in some unit of measure, as contrasted with a percentage or ratio of that amount to the total of its data series; quantity.

anomalous Statistics jargon for describing results that defy explanation.

approximation In mathematics, an inexact value; data reduction by summary, rounding, or truncation.

area In geometry, the product of a rectangular object's height and width; a chart type in which the portion beneath an *xy* line plot is shown as a solid object.

argument In the *syntax* of a spreadsheet function, a reference to a data item, such as a cell address or range name, that will be processed by the function.

ascender In typesetting, the portion of a lowercase letter that extends upward from its body, or main portion.

attribute A characteristic, such as color, of a font or graphic object.

average A kind of data reduction produced by summing a series of numbers, and then dividing by the count (the number of items in the series).

axis (plural *axes,* pronounced *ak*-seez) In an *xy* or a radar chart, a dimension representing a measurement or criterion according to which one coordinate of the data will be plotted; see also *Cartesian coordinate system, scale.*

B

bar An *xy* chart format in which data points are indicated by the heights of separate graphic columns, or bars.

bar-line　An *xy*-chart plotting style which plots some data series as bars, and others as lines.

baseline　The reference line in a chart from which distances of the *y* coordinates of data points are measured; in an *xy* chart with no stacking, the line *y*=0; in a stacked chart, the baseline is formed by the plot of the preceding data series; in typesetting, the line on which letters sit.

bitmap　In computer graphics terminology, an array of colored dots; a picture.

bound　Maximum or minimum value on a chart axis or data series.

bubble　A chart type for showing the relationship of *sets* against time and complexity or some other pair of *xy* scales.

build-up sequence　Division of an on-screen text list into a series of screens so that the list is revealed one topic at a time, as covered in the presenter's speech.

C

callout　In a chart, a line or arrow that connects a chart object such as a pie slice with its descriptive label.

Cartesian coordinate system　In trigonometry and analytic geometry, a scheme of charting (named for French mathematician and philosopher René Descartes) that locates points in physical space according to three perpendicular dimensions, or axes: width (*x*), height (*y*), and depth (*z*).

category　A set of criteria by which data can be sorted; in mathematics, the equivalent of a *set*; in a computer database, the equivalent of a *field*.

cell　One address in a spreadsheet, or the intersection of a column and a row, which can hold just one data item.

character set　The collection of letters, punctuation marks, and symbols in a specific font.

chart　A graphic that shows one or more data series as objects; a graphic screen or page in a presentation.

chart type　A principal charting format: *xy*, pie, radar, Venn, bubble; see also *plotting style*.

chroma　Color purity; see also *tint*.

circular reference　In a spreadsheet, any formula that refers to itself, whether directly or indirectly.

clip art　Ready-made artwork sold for use in graphic materials such as presentations, typically on digital computer media such as CD-ROM.

cluster　In a bar chart, a style in which data series are shown as separate bars, grouped at major scale divisions.

CMY or **CMYK**　A system of defining color in printed media such as magazines as a subtractive combination of cyan, magenta, yellow, and black; four-color printing process.

color　An attribute of the way objects reflect light, an element of visual design; a specific color *tone*, a combination of hue, tint, and shade.

color film recorder　A computer output device that creates images on photographic slide film.

color gamut　A range of color tones that can be reproduced by a specific device, such as a video monitor or a computer printer.

color palette   A range of color tones in a computer graphics program from which a designer can pick tones.

color scheme   A designer's selection of color tones for use in a presentation.

color wheel   A circular arrangement of *RGB* and *CMY* colors used by graphic designers in making color selections.

column   In a pie chart, a second data series shown as a stacked bar, which itemizes one of the pie slices; in a table, all of the positions (cells) under one heading, or field name.

composition   In graphic design, the arrangement of graphic objects on a page or screen.

compounding   In a financial investment, periodic combination, or reinvestment, of interest with principal so that interest in the succeeding period is calculated on the reinvested total.

constant   In mathematics, a fixed amount; not variable; see also *scale factor.*

contrast range   A difference in light intensity between the lightest and darkest color tones in a picture.

coordinate   One of two or three values (*x*, *y*, or possibly *z*) with which a data point will be plotted in an *xy* chart, based on the Cartesian coordinate system.

correlation   In mathematics, the causative link between *dependent* and *independent variables.*

count   The number of items in a data series, which can be distorted depending on whether the items include zeros or blanks.

credit line   An explanatory note on a chart or photograph that indicates its source.

cross-foot   An accounting technique for verifying the accuracy of a total: adding both down the columns and across the rows of a numeric table should produce the same *grand total.*

CRT   A cathode ray tube; a display device in a television or computer monitor.

cumulative   A plot in which each point is a running total of all preceding points in the data series; *not* a synonym for *stacked!*

curve   In charting, any trend line, whether straight or curved.

custom colors   A set of color tones not contained in a predefined color palette; a mixed color.

D

data   The plural of the Latin *datum*, meaning *fact;* often used as a singular noun to refer to a collection of data items; a series of data items plotted in a chart or shown in a table.

data item   One point or value in a data series; in mathematics, a member of a set.

data processing   The means by which raw data is transformed into useful information.

data reduction   Using arithmetic to summarize a detailed set of data, with the purpose of making it simpler and easier to understand, often resulting in a single number.

data series or **data set**   A collection of numeric measurements plotted on a chart.

data table   In a chart, a supplementary display of data values, grouped by columns and rows, often incorporating the chart *legend.*

data type   A restriction on a category of data item requiring entries to be alphabetic, numeric, alphanumeric (text, also called *string*), logical (true or false), or in the specific numeric form of an integer, a decimal, and so on.

database   A specific format for tables in which each column heading is a field name for just one data type, such as a phone number, and each row is a record describing a single entity, such as a person.

delimiter   A punctuation or mark that indicates the division between data *fields* in a label or in a data record.

delta   (written as the Greek symbol Δ)   In mathematics, the difference between two values: $y_1 - y_2 = \Delta y$; see also *slope*.

dependent variable   In mathematics, a value that necessarily changes when some other value under consideration changes.

descender   In typesetting, the portion of a lowercase letter that extends below the baseline.

diagram   A drawing that shows things or concepts as linked symbols; a schematic drawing.

diameter   The full height or width of a circle; in geometry, any chord of a circle that passes through its center.

dimension   A direction in physical space; in charting, one axis of an xy chart.

dithering   On computer displays, the juxtaposition of two dots of different colors to give the impression of a third color.

division   One marked unit on the scale of an xy chart axis.

donut   A charting format that resembles a set of concentric, hollowed-out pie charts.

dual-y axes   Characteristic of an xy chart that has two separate y-axis scales, usually left and right, for plotting data series measured in different units on the same chart.

duration   A span of time; in a Gantt chart or project schedule, the length of a task or activity.

E

ellipsis   In typesetting, three dots (...) used to indicate an omitted word or phrase.

em dash, en dash   In typesetting, the long dash (width of the letter M) and the short dash (width of the letter n) in a font.

entity   In data processing, a person, organization, or thing described in a database record.

error bar   In xy charting, a graphic technique for indicating a range of uncertainty or deviation.

explode   To pull one or more slices of a pie chart outward from its center for emphasis.

exponent   A power by which a number is raised, as indicated by a superscript: $2^3 = 2 \times 2 \times 2 = 8$.

exponential   A data series that increases by a power, or *exponent*; increasing by *orders of magnitude*.

extrapolation   Projecting future values from a current series of data.

F

fact checking   In report writing, the investigation of cited facts for accuracy by reading reference materials or by contacting witnesses or authorities.

fair use   A legal term for quotation or reproduction of small portions of copyrighted material without formal permission from the copyright holder.

field   In a *database* table, a heading that describes the data entries in a column; an entry in a data record, or row, of a database.

floating bar   A bar that does not begin, or have its minimum y value, at the chart baseline.

font   A typeface in a specific size; generally, a typeface irrespective of size.

foot   An accounting term for adding, or summarizing, a column of numbers.

formula   A math problem that can be contained in a cell of a computer spreadsheet, referring to values that are held in other cells in the sheet.

four-color process   A printing technology that uses the *CYMK* system of color.

function   In spreadsheet terminology, a formula that is predefined in the program.

G

Gantt chart   A scheduling chart with time spans, or task *durations*, plotted as horizontal bars; named for industrial engineer Henry Gantt.

grand total   In an accountant's ledger, the one total that summarizes all the other totals in a table.

graph   An *xy* chart; any chart that shows one or more data series as plots or curves.

grayscale   A range of monochrome shades.

H

handicap   In golf, a clever technique of scoring that compensates for differences in the skill levels of competing players; a number of strokes by which a golfer is permitted to subtract from his or her score; a shameless liar's trick.

high-low-close-open (HLCO)   A special *xy* chart format for plotting stock prices.

horizontal orientation   An exceptional form of *xy* chart in which the *x* axis is vertical, the *y* horizontal; see also *vertical orientation*.

HSL   A system used by graphic artists to describe color tones in terms of hue, saturation, and lightness.

hue   One of the primary colors from which color tones are blended.

I

independent   In mathematics, a variable that is not necessarily affected if the value of some other variable under consideration changes.

intensity   Another term for color *shade*.

international characters   Accented letters and symbols contained in a font for composing text in Romance languages.

interpolation   Estimating the missing, or in-between, values in a data series.

intersection   The values or data items that two *sets* hold in common.

iteration   In spreadsheet terminology, the repetition of a calculation (also called *pass* or *loop*).

J

Jones   The guy who wrote this book; American slang for an irrepressible urge.

K

key   See *legend*.

keyword   In spreadsheet terminology, a reserved name used as an argument to request a specific type of result from a *function*.

L

label   Text displayed in a chart that names or explains a displayed value or plot; in spreadsheet terminology, any text entry.

legend In a chart, a key that correlates the names of data series to their plots, usually by color codes or patterns.

letterform In typesetting, the design of a character in a font.

lightness Another term for color *shade*.

line An *xy* chart format in which data points are connected to form a continuous, segmented line (polyline), or *line plot*; a geometric object defined as the shortest distance between two separate points in a plane.

linear Describes a straight line, which is the shortest distance between two points; an *xy* chart that is not logarithmic; see also *logarithmic*.

linked bars An *xy*-chart plotting style in which tops of bars are connected by line segments to emphasize trends.

list An unstructured collection of rows that contain dissimilar but perhaps related data records, as contrasted with the structure of rows in a table or database.

logarithmic Refers to an *xy* chart in which one axis (*semi-log*) or both axes (*log-log*) have been scaled by powers of 10; see also *scaling*.

log-log An *xy* chart in which both axes have been scaled by powers of 10.

M

major division A tick on a chart axis scale that marks the primary increment of its unit of measure; in all types of *xy* charts except scatter charts, the ticks on the *x* axis at which data items are plotted in sequence—from left to right in vertical charts, from top to bottom in horizontal charts.

mean The average of just the high and low values in a data series; the mid-point.

median In a data series, the item located exactly in the middle of the count of items, where there are the same number of values above it as are below it; if there are an even number of items in the series, the median is the average of the middle two items.

mid-point See *mean*.

minor division A tick on a chart axis scale that marks the secondary increment of its unit of measure; the subdivision of a major tick interval on a chart scale.

monospace In typesetting, a font in which the spacing between letters is uniform.

mood The emotional response of an audience, which can be affected by on-screen colors.

moving average In an ongoing series of values, the arithmetic average of only a specific number of recent ones; for example, in a lifetime of golf scores, the most recent four-game average.

multivariate analysis An investigation of several factors or criteria that contribute to a single result or that describe a single entity (one purpose of radar charting).

N

nesting In mathematics and in computer programming, refers to a process, such as an equation or formula, that is contained within a larger process; use of sets of parentheses in algebra or in spreadsheet formula syntax to control the order of processing.

O

object In computer graphics, a basic drawing element, such as a string of text or a geometric shape; a collection of graphic objects that can be manipulated as a group.

order of magnitude A power of 10; for an integer or decimal value, the number of digits to the left of the decimal point, minus one— for example, the order of magnitude of 160,000 is 5 (1.6×10^5).

ordinal Describes a data series for which the sequence of items uses some method of counting: 1, 2, 3…or January, February, March…

orientation The position of a chart element in relation to something else; with respect to the chart frame, the *y*-axis options *vertical* or *horizontal;* in a presentation, the way in which a rectangular sheet of printed paper or a display screen is to be viewed—*portrait* (long dimension vertical) or *landscape* (long dimension horizontal); in graphic design, the position and direction of graphic shapes in relation to a page or screen; see also *composition.*

origin In the 2D Cartesian coordinate system, the point at which *x*=0 and *y*=0, or (0,0); in a vertical *xy* chart in which both scales start at 0, the lower-left corner of the axes.

overlay A clear sheet used by graphic artists as a separate drawing layer; in a 2½D Cartesian coordinate system, a drawing plane.

P

paired-bar A horizontal *xy* plotting style in which two comparable data series are shown as opposed sets of bars—one plotted from the center rightward, the other from the center leftward.

perspective A two-dimensional drawing technique used to render the third dimension, or depth, as lines that converge on a distant point on the horizon; see *vanishing point.*

pica A typesetter's unit of spacing, equal to ⅙ inch.

pie A circular or elliptical chart in which percentages of the whole are shown as slices, or sectors.

plagiarism The unauthorized and uncredited use of material from a copyrighted source; see also *substantial similarity.*

planar Having the characteristics of a geometric plane; two-dimensional; flat.

plot In an *xy* chart, the line or curve described by connecting the points in a data series.

plotting style A set of options within a chart type: bar, stacked bar, bar-line; see also *chart type.*

PMS colors Color tones as defined by the Pantone Matching System developed by Pantone, Inc. (Carlstadt, NJ).

point The location of a data value on a chart; on a vertical *xy* chart, a point is defined by a horizontal distance (*x*) and a vertical distance (*y*) from the origin (0,0); the delimiter in a decimal value; a typesetter's unit of measure equal to ¹⁄₇₂ inch; see also *scatter.*

POM dependency Sardonic slang term used by computer programmers to describe results that seem to have no apparent, consistent, or controllable cause—therefore, depending on the phase of the moon.

population In statistics, a *standard deviation* or other measurement that includes all items in a data set.

precision In a decimal value, the number of digits to the right of the decimal point.

primary colors In the *RGB* color wheel, pure hues such as red, blue, and yellow.

progress In the graphics of Western cultures, movement from left to right.

proportional   In mathematics, refers to the ratio of one value to another, describing a percentage; in typesetting, a font that permits variable spacing between letters.

public domain   A legal term describing material for which there is no private copyright holder.

Q

quantity   A numeric value that requires a unit of measure for a label; a desirable feature if you also have *quality*; see also *amount*.

R

radar   A chart format for investigating symmetrical relationships in which axes for plotting multiple criteria radiate from the center like the spokes of a wheel; what they got me with on the road between Phoenix and Payson.

range   The extent of a chart axis or data series that lies between its bounds, or minimum and maximum values; a data series; a set of permissible values; in a spreadsheet, a contiguous block of cells, each of which can contain one data item.

readability index   A score that indicates the educational grade or skill level required to read and comprehend a specific document.

reasonableness   An accounting term for the appearance of accuracy in results.

record   In a *database* table, one row, which must describe just one entity, such as a person. (A row *must* describe one and only one entity.)

regression   In mathematics and charting, a technique for fitting a smooth curve through a given set of data points; what will happen to you if you work too hard writing books; technically, a formula in calculus for minimizing the residual square error between data points and the fitted curve in a process called *least squares analysis*.

relative   A value not expressed as an explicit number of some unit of measure, but as measured against some other value of the same data type; for example, a ratio.

RGB   A system used in color computers and video cameras for describing color tones as mixtures of red, green, and blue.

rotate   To adjust the starting point of a pie chart by some offset from zero degrees, usually 0–360 degrees counterclockwise.

rounding   The approximation of a long decimal value to a fewer number of decimal places by dropping all but a given number of digits to the right of the decimal point, possibly adjusting the right-most remaining digit to approximate the discarded digits; if the right-most digit is increased, the process is called *rounding up*, if decreased, *rounding down*.

S

sample   In statistics, a *standard deviation* or other measurement that includes a selected portion of all items in a data set.

sans serif   In typesetting, a font that has no ornamentation at the ends of its letters, such as Arial and Helvetica.

saturation   Another term for color *chroma* or *tint*.

scalable   Refers to a font that can be adjusted continuously in size, rather than being restricted to discrete point sizes.

scale   A chart axis marked at regular intervals, or *divisions*, according to a unit of measure.

scale factor   A constant by which a chart axis scale is multiplied or divided, such as thousands or tenths.

scaling   Multiplying or dividing chart axis scale values by a constant, or fixed, amount; see also *scale factor.*

scatter   A type of *xy* chart in which *x* values need not be plotted in sequence at major scale divisions, removing any association between the *x* axis and the progress of time; a point chart.

secondary colors   In the *RGB* color wheel, hues that are blends of primary colors, such as red-violet.

secondary y axis   In an *xy* chart, an optional y_2 axis; see *dual-y axes.*

segmented bar   See *stacked.*

semi-log   An *xy* chart in which one axis, usually the *y*, has been scaled by powers of 10.

sense   In grammar, the correct interpretation of a written sentence, which should be clear from its composition and syntax.

serial comma   In English punctuation, the use of a comma to separate words listed within a sentence.

serial date   A decimal form of calendar notation, including the date and time, used in computer processing; the digits to the left of the decimal point represent the date, counting from the year 1900 (or 1904 on some systems), and the digits to the right represent the 24-hour time in hours, minutes, and seconds.

serif   A font with cute doodads and curlicues at the ends of its precious letterforms; one such doodad or curlicue.

set   In mathematics, a group of related things, usually numbers; a data series.

shade   The amount of black in a color tone.

shape   The outline of a graphic object; see also *curve.*

sheet   Jargon for *spreadsheet* or *worksheet.*

signposting   A speechwriter's technique for listing the major topics of a presentation on the screen.

skew   A distortion of the result of a data reduction due to unusually high or low values in the data series.

skip factor   A specification for omitting selective labels from the scale of a chart axis; for example, a skip factor of five would cause every fifth scale label to be displayed on the axis.

slice   A data item, shown as a percentage, in a pie chart; also called a *cut, piece, segment, wedge* or *sector.*

slip card   A piece of cardboard used to hide upcoming presentation topics listed on an overhead transparency.

slippage   In a Gantt chart, the visible gap between a task's estimated start date and the actual start date.

slope   A slant of a chart line or curve, defined for two data points by the change in *y* (Δy) divided by the change in *x* (Δx); the rate of change.

stacked   A charting option for showing data series in bars or areas that are built on top of one another, like a stack of bricks.

standard deviation   In statistics, the degree to which a data value differs from the *mean.*

style sheet   A type of *template* used to create documents; a collection of graphic options used to define the appearance of a document.

subscript   In mathematics, a notation used to indicate the sequential number of a data item within a series: x_3; the screenplay for *The Enemy Below.*

substantial similarity   The legal term used to define material that has been plagiarized, or used without permission from a copyrighted source.

subtotal   In a table, a component of a *grand total*.

subtractive color   The result of filtering reflected light through colored dyes on different layers of photographic film or through printing inks overlaid on paper.

subtype   Variations within a chart type: area, stacked area, 3D area; see also *plotting style*.

summary   A type of data reduction produced by adding a series of numbers; a total.

syntax   In spreadsheet terminology, the required written form of a formula or function; in English composition, the structure of a sentence.

T

template   In computer graphics, a generic design that lacks only the user's data to create a finished chart.

thematic coding   The use of color to identify principal topics in a presentation.

three-dimensional (3D)   Chart plotting styles such as lines, bars, or areas shown with an edge to suggest thickness or depth.

tick   A short line segment; a mark on a chart scale that indicates a major or minor division; a mark on a floating bar in an HLCO chart that indicates the opening or the closing price; in financial trading, the difference between the opening and closing prices on a given day.

time   In a typical *xy* chart, the *x* coordinate, which varies along a left-to-right (horizontal) *x* axis; in a Gantt chart, the span of a horizontal bar.

time analysis   The scheduling process in which the starting and ending dates of tasks are correlated to find the *critical path*, or minimum time span required to complete all essential tasks.

tint   The amount of white in a color tone.

trend   The identifiable pattern in a series of data items; a chart curve that describes this pattern from which future results can be *extrapolated* and intermediate results can be *interpolated*.

trend analysis   The study of data or plots in charts in an attempt to find meaningful patterns.

truncation   The shortening of a long decimal value to fewer decimal places simply by dropping all but a given number of digits to the right of the decimal point; see also *rounding*.

two-and-a-half-dimensional (2½D)   Cartesian coordinate system for plotting *xy* charts in which a third axis, *z*, is used to fix the position of multiple drawing planes, or overlay sheets.

typeface or **face**   A style of lettering; a font, irrespective of its point size.

typesetting   The reproduction of text in print, according to a specific graphic design.

U

union   In mathematics and logic, the combination of two *sets*; the marriage of like minds (which admits no impediment).

V

value   A number in a data series; another term for color *shade;* see also *label*.

vanishing point   In a two-dimensional drawing that renders the third dimension by using perspective, a point on the horizon toward which depth lines converge.

variable   In mathematics, a placeholder for an unknown value or any of a range of possible values.

Venn   A type of diagram, named for English logician John Venn, that uses circles or ellipses to show relationships between sets and subsets.

vertical orientation   A typical form of xy chart in which the y axis is vertical, the x horizontal; see also *horizontal orientation*.

visual bias   A set of assumptions shared by members of a cultural or professional group about the inherent meanings of graphic compositions and elements.

visual cueing   The appearance of text or an object on the screen at the same moment it is spoken or named by the presenter; television newswriters often screw this up because, working against tight deadlines, they have no idea what picture the producer will find to put with their words!

visual priority   In 2½D or 3D Cartesian coordinate systems, the z-axis value of a graphic object, which determines whether it will cover or be hidden by other objects; objects with low priority (high z values) will be in the background and will be hidden where they overlap with objects of high priority (low z values), which will be in the foreground.

volume   Total amount (volume of financial trading); 3D spatial measurement (height times width times depth).

W

weight   In statistics, a factor or multiplier used to compensate for an incompatibility between data sets, such as different units of measure or levels of importance.

workpaper   An accounting term for an interim document; scratchpad.

workstation   A desktop computer or terminal; technically, a powerful microcomputer used for graphic or engineering applications.

X

x axis   In a typical xy chart, the horizontal scale against which the first coordinate x (usually representing a point in time) is plotted.

xy   A chart format based on the 2D Cartesian coordinate system in which time typically progresses along the horizontal x axis and measured quantities vary along the vertical y axis.

Y

y axis   In a typical xy chart, the vertical scale against which the second coordinate y (quantity) is plotted.

Z

z axis   In the 3D Cartesian coordinate system for xy charts, a third axis that indicates depth; in a 2½D system, the position of the drawing planes, or overlay sheets.

zzzzzzzz   A soporiferous condition not likely to be induced by this stimulating discourse on an otherwise dreary motif.

Index

Note to the Reader: Throughout this index, **boldfaced** page numbers indicate primary discussions of a topic. *Italicized* page numbers indicate illustrations.

FOR EVERY COMPUTER QUESTION,
THERE IS A SYBEX BOOK THAT HAS THE ANSWER

Each computer user learns in a different way. Some need thorough, methodical explanations, while others are too busy for details. At Sybex we bring nearly 20 years of experience to developing the book that's right for you. Whatever your needs, we can help you get the most from your software and hardware, at a pace that's comfortable for you.

We start beginners out right. You will learn by seeing and doing with our **Quick & Easy** series: friendly, colorful guidebooks with screen-by-screen illustrations. For hardware novices, the **Your First** series offers valuable purchasing advice and installation support.

Often recognized for excellence in national book reviews, our **Mastering** titles are designed for the intermediate to advanced user, without leaving the beginner behind. A **Mastering** book provides the most detailed reference available. Add our pocket-sized **Instant Reference** titles for a complete guidance system. Programmers will find that the new **Developer's Handbook** series provides a more advanced perspective on developing innovative and original code.

With the breathtaking advances common in computing today comes an ever increasing demand to remain technologically up-to-date. In many of our books, we provide the added value of software, on disks or CDs. Sybex remains your source for information on software development, operating systems, networking, and every kind of desktop application. We even have books for kids. Sybex can help smooth your travels on the **Internet** and provide **Strategies and Secrets** to your favorite computer games.

As you read this book, take note of its quality. Sybex publishes books written by experts—authors chosen for their extensive topical knowledge. In fact, many are professionals working in the computer software field. In addition, each manuscript is thoroughly reviewed by our technical, editorial, and production personnel for accuracy and ease-of-use before you ever see it—our guarantee that you'll buy a quality Sybex book every time.

To manage your hardware headaches and optimize your software potential, ask for a Sybex book.

FOR MORE INFORMATION, PLEASE CONTACT:

Sybex Inc.
2021 Challenger Drive
Alameda, CA 94501
Tel: (510) 523-8233 • (800) 227-2346
Fax: (510) 523-2373

SYBEX

Sybex is committed to using natural resources wisely to preserve and improve our environment. As a leader in the computer books publishing industry, we are aware that over 40% of America's solid waste is paper. This is why we have been printing our books on recycled paper since 1982.

This year our use of recycled paper will result in the saving of more than 153,000 trees. We will lower air pollution effluents by 54,000 pounds, save 6,300,000 gallons of water, and reduce landfill by 27,000 cubic yards.

In choosing a Sybex book you are not only making a choice for the best in skills and information, you are also choosing to enhance the quality of life for all of us.

GET A FREE CATALOG JUST FOR EXPRESSING YOUR OPINION.

Help us improve our books and get a *FREE* full-color catalog in the bargain. Please complete this form, pull out this page and send it in today. The address is on the reverse side.

Name _____   **Company** _____

Address _____   **City** _____ **State** ____ **Zip** _____

Phone ( )_____

1. How would you rate the overall quality of this book?

❑ Excellent
❑ Very Good
❑ Good
❑ Fair
❑ Below Average
❑ Poor

2. What were the things you liked most about the book? (Check all that apply)

❑ Pace
❑ Format
❑ Writing Style
❑ Examples
❑ Table of Contents
❑ Index
❑ Price
❑ Illustrations
❑ Type Style
❑ Cover
❑ Depth of Coverage
❑ Fast Track Notes

3. What were the things you liked *least* about the book? (Check all that apply)

❑ Pace
❑ Format
❑ Writing Style
❑ Examples
❑ Table of Contents
❑ Index
❑ Price
❑ Illustrations
❑ Type Style
❑ Cover
❑ Depth of Coverage
❑ Fast Track Notes

4. Where did you buy this book?

❑ Bookstore chain
❑ Small independent bookstore
❑ Computer store
❑ Wholesale club
❑ College bookstore
❑ Technical bookstore
❑ Other _____

5. How did you decide to buy this particular book?

❑ Recommended by friend
❑ Recommended by store personnel
❑ Author's reputation
❑ Sybex's reputation
❑ Read book review in _____
❑ Other _____

6. How did you pay for this book?

❑ Used own funds
❑ Reimbursed by company
❑ Received book as a gift

7. What is your level of experience with the subject covered in this book?

❑ Beginner
❑ Intermediate
❑ Advanced

8. How long have you been using a computer?

years _____
months _____

9. Where do you most often use your computer?

❑ Home
❑ Work

❑ Both
❑ Other _____

10. What kind of computer equipment do you have? (Check all that apply)

❑ PC Compatible Desktop Computer
❑ PC Compatible Laptop Computer
❑ Apple/Mac Computer
❑ Apple/Mac Laptop Computer
❑ CD ROM
❑ Fax Modem
❑ Data Modem
❑ Scanner
❑ Sound Card
❑ Other _____

11. What other kinds of software packages do you ordinarily use?

❑ Accounting
❑ Databases
❑ Networks
❑ Apple/Mac
❑ Desktop Publishing
❑ Spreadsheets
❑ CAD
❑ Games
❑ Word Processing
❑ Communications
❑ Money Management
❑ Other _____

12. What operating systems do you ordinarily use?

❑ DOS
❑ OS/2
❑ Windows
❑ Apple/Mac
❑ Windows NT
❑ Other _____

13. On what computer-related subject(s) would you like to see more books?

_____

_____

_____

_____

14. Do you have any other comments about this book? (Please feel free to use a separate piece of paper if you need more room)

_____

_____

_____

_____

- - - - - - - - - - - PLEASE FOLD, SEAL, AND MAIL TO SYBEX - - - - - - - - - -

SYBEX INC.
Department M
2021 Challenger Drive
Alameda, CA
94501

SYBEX®

Let us hear from you.

 Talk to SYBEX authors, editors and fellow forum members.

 Get tips, hints and advice online.

 Download magazine articles, book art, and shareware.

Join the SYBEX Forum on CompuServe®

you're already a CompuServe user, just type **GO SYBEX** to join the
YBEX Forum. If not, try CompuServe for free by calling 1-800-848-8199
nd ask for Representative 560. You'll get one free month of basic
ervice and a $15 credit for CompuServe extended services—a $23.95
alue. Your personal ID number and password will be activated when
ou sign up.

Join us online today. Type GO SYBEX on CompuServe.
If you're not a CompuServe member, call Representative 560
at 1-800-848-8199 .

SYBEX

(outside U.S./Canada call 614-457-0802)